A Field Guide to Insects in Australia

A Field Guide to Insects in Australia

Green lacewing or golden-eye, **Chrysopidae** (25mm long).

SECOND EDITION

Paul Zborowski and Ross Storey

Reed New Holland

Published in Australia by
Reed New Holland
an imprint of New Holland Publishers (Australia) Pty Ltd
Sydney • Auckland • London • Cape Town

14 Aquatic Drive Frenchs Forest NSW 2086 Australia
218 Lake Road Northcote Auckland New Zealand
86 Edgware Road London W2 2EA United Kingdom
80 McKenzie Street Cape Town 8001 South Africa

Published in 1995 by Reed Books
Reprinted in 1996 and 1997
Reprinted in 1998, 2000 and 2002 by Reed New Holland
Second edition published 2003

Copyright © 1995, 2003 in text: Paul Zborowski and Ross Storey
Copyright © 1995, 2003 in photographs: Paul Zborowski or as credited
Copyright © 2003 New Holland Publishers (Australia) Pty Ltd

All rights reserved. No part of this publication may be reproduced,
stored in a retrieval system or transmitted, in any form or by any means,
electronic, mechanical, photocopying, recording or otherwise, without the
prior written permission of the publishers and copyright holders.

National Library of Australia Cataloguing-in-Publication Data:

Zborowski, Paul
A field guide to insects in Australia

2nd ed.
Includes index
ISBN 1 87633 496 7

1. Insects—Australia. 2. Insects—Australia—Identification.
3. Insects—Australia—Anatomy. 4. Insects—Australia—Geographical distribution.
I. Storey, Ross. II. Title.

595.70994

Publisher: Louise Egerton
Project Editor: Yani Silvana
Editor: Sharon Paull
Designer: Trevor Hood/Anaconda Graphic Design
Printer: Hong Kong Graphics

FRONT COVER
TOP LEFT: Stag beetle, *Lamprima latreillii*, **Lucanidae**. TOP CENTRE: Flower wasp, *Thynnus pulchralis*, **Tiphiidae**. TOP RIGHT: Northern green grocer cicada, *Cyclochila virens*, **Cicadidae**. CENTRE: Queensland day moth, *Alcides zodiaca*, **Uraniidae**. BOTTOM: Praying mantis, **Mantidae**.

BACK COVER
TOP LEFT: Red dragon fly, *Diplacodes bipunctata*, **Libellulidae**. BOTTOM LEFT: Fruitpiercing moth caterpillar, *Othreis fullonia*, **Noctuidae**. TOP RIGHT: Hercules moth, *Coscinocera hercules*, **Saturniidae**. BOTTOM RIGHT: Blue-banded bee, *Amegilla* sp. **Anthophoridae**.

ACKNOWLEDGMENTS

It is much easier to identify an insect in the hand than from a photograph and so the difficult task of naming the species illustrated here was shared among many specialists.

We are indebted to those listed below for their efforts (any errors of identification that may still be present in this book are entirely due to the final discretion of the authors): John Balderson, Jo Cardale, Mary Carver, Gerry Cassis, Don Colless, Ted Edwards, Marianne Horak, Ken Key, Kevin Lambkin, John Landy, David McAlpine, Leigh Miller, Max Moulds, Ian Naumann, Tim New, David Rentz, Curtis Reid, Steven Shattuck, Courtney Smithers, John Trudinger, Alice Wells, Tom Weir and David Yeates.

Special thanks to Rob and Ruth Whiston for their help in some difficult photographic assignments; Ceri Pearce for test-driving the key to orders; and David Rentz, Geoff Monteith, CSIRO Entomology, and Geoff Thompson and Robyn Redman for providing extra illustrations.

Bull ant, *Myrmecia* sp., **Formicidae** (22 mm long)

CONTENTS

Acknowledgments 5
Preface to the Second Edition 8
Introduction 8

PART 1
What Is an Insect? 10
Insect Life Cycles 16
Crypsis and Mimicry 19
Collecting Insects 24
Classification and a Key to the Insect Orders 29

PART 2
The Pseudo Insects
   Springtails, Proturans, Diplurans (The Entognatha) 39

Class Insecta: The Apterygota
   Bristletails (Order Archaeognatha) and Silverfish
     (Order Thysanura) 42

Class Insecta: The Pterygota
   Mayflies (Order Ephemeroptera) 44
   Dragonflies and Damselflies (Order Odonata) 46
   Stoneflies (Order Plecoptera) 50
   Cockroaches (Order Blattodea) 52
   Termites (Order Isoptera) 55
   Praying Mantids (Order Mantodea) 59
   Earwigs (Order Dermaptera) 62
   Crickets and Grasshoppers (Order Orthoptera) 64
   Stick Insects and Leaf Insects (Order Phasmatodea) 73
   Web Spinners or Embiids (Order Embioptera) 77
   Booklice and Psocids (Order Psocoptera) 78

Antlion, **Myrmeleontidae** (45 mm long)

Lice (Order Phthiraptera)	80
True Bugs, Hoppers, Scale Insects and Aphids	
(Order Hemiptera)	82
Thrips (Order Thysanoptera)	97
Alderflies and Dobsonflies (Order Megaloptera)	99
Lacewings, Antlions, Mantis Flies	
(Order Neuroptera)	101
Beetles (Order Coleoptera)	108
Stylopids (Order Strepsiptera)	129
Scorpion Flies and Hanging Flies (Order Mecoptera)	131
Fleas (Order Siphonaptera)	133
Flies (Order Diptera)	135
Caddisflies (Order Trichoptera)	151
Moths and Butterflies (Order Lepidoptera)	155
Wasps, Ants, Bees and Sawflies (Order Hymenoptera)	179
Glossary	199
Multimedia Bibliography	203
Index	204

PREFACE TO THE SECOND EDITION

This new edition of *A Field Guide to Insects in Australia* has been made more accessible by increasing the use of common names and making them more prominent. Sixteen new colour photographs have been added. Information and scientific names have been updated where necessary. New, up-to-date references have been added to the chapters where possible. In addition, a Multimedia Bibliography, including books, CD-roms and web sites, has been added.

INTRODUCTION

With over a million species worldwide and a historical and not always beneficial association with human endeavour, insects are very much a part of our lives. Much has been written about insects, especially the economically important ones; however, most of the insect species have only initial descriptions, or are still awaiting formal description. This sorting and naming of organisms is the work of taxonomists and this book is an introductory taxonomic guide. It is aimed at users who have little or no knowledge of insects, but who have a curiosity about their amazing diversity.

How To Use this Guide

The primary objective of this guide is to provide a means to identify an insect in the hand to the level of order, with reasonable accuracy and with minimal prior knowledge.

The book is divided into two parts. Part 1 provides the background information on insect appearance, life cycles and classification. To be able to identify an insect, it is essential that you first study the introductory anatomy chapter, 'What Is an Insect?'. There are hundreds of body parts in the insect exoskeleton and sometimes the same parts

INTRODUCTION

have different names when applied to different insect groups. The adjectives that describe the shape of these body parts form a large specialist dictionary in themselves. This guide attempts to minimise this potential confusion by referring to a limited number of body parts and by using illustrations to explain the specialist descriptive terms. For instance, the shape of the antennae is an important identification characteristic and at least 10 words exist to describe these shapes. Any terms used in this guide are either first illustrated in 'What Is an Insect?' or defined in the Glossary.

Also in Part 1 is the 'Classification and Key to Insect Orders' (page 29) which describes the principles of classification and has a key that guides the user through a series of questions which help to determine in which order an insect belongs.

In this book we follow the classification and current information on species numbers as presented in the most comprehensive insect text to date, *The Insects of Australia* (1991), published by the CSIRO. For the common names used here, we refer to Australian Insect Common Names at http://www.ento.csiro.au/aicn/.

Part 2 of the guide consists of a chapter for each order found in Australia. These chapters give a brief description of the characteristics which in combination define each order, followed by general information on the life cycle and biology of typical species. Each chapter also has a classification section which introduces major families of each order and lists distinctive characteristics that define these families and thus provide the means to determine if a specimen may belong to one of these. Most of the described families are illustrated with colour photographs of one or more species as they appear alive in the field. There is a trend in insect guide books to show paintings or photographs of spread, pinned insects. This is excellent for discerning fine detail, but it does not show the insect in the way you will see it in life. The ways in which a live insect stands at rest, feeds or moves are often good diagnostic features and this guide includes these field characteristics whenever possible.

If you wish to pursue the search further, every chapter has a recommended reading list for books and sometimes journal articles that give more detailed descriptions. Many of these sources have no glossaries and therefore will be more difficult to use than this guide. However, the experience of using this guide and glossary will serve as a good introduction to them.

WHAT IS AN INSECT?

To demonstrate what an insect is, it is first necessary to step back and look at the larger picture of which insects are a part. The Animal Kingdom is divided into several large groups called phyla. At this level, all backboned animals, for example, are in one phylum, Chordata. In terms of species numbers this is a small group. By far the largest is the phylum Arthropoda which includes the insects. This group is characterised by having an external skeleton called an exoskeleton which is composed of separate hardened plates or segments joined by softer tissue that allows movement of these segments. This structure gives strong protection from the environment on the outside and sturdy points for muscle attachment on the inside. It is such a successful formula that arthropods represent close to 90% of all animal species, with the insects accounting for the majority of these.

Among the many classes in Arthropoda, the insects (class Insecta) are characterised by having three pairs of legs, one pair of antennae, generally three obvious body divisions, and by being the only arthropods to possess wings. These are adult characteristics and are not always obvious. Other arthropod classes with similar features have enough differences to avoid confusion with insects in most instances (see Figure 1):

- Arachnida (spiders, scorpions, ticks and mites): only two body divisions as the head is incorporated into the second division; four pairs of legs; and antennae and wings are absent.
- Myriapoda (centipedes and millipedes): many pairs of legs, never less than nine pairs; one pair of antennae;

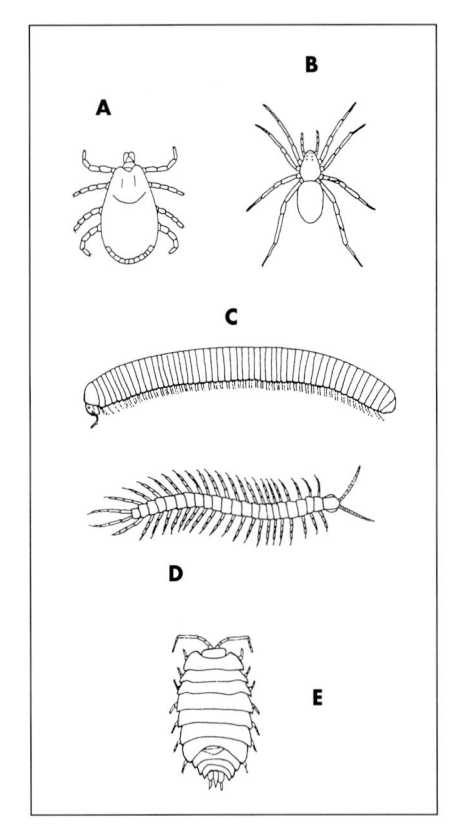

Figure 1. Various non-insects. Arachnida: (a) tick; (b) spider. Myriapoda: (c) millipede; (d) centipede. Crustacea: (e) woodlouse (Isopod).

many body segments, all in one elongate body shape; and wings are absent.
- Crustacea (including crabs, shrimps, woodlice): several to many pairs of legs, never less than four; two pairs of antennae; body shape is variable, never with three main divisions; wings are absent.

External Anatomy of Insects

Adult insects have a general body plan of three main divisions: head, thorax and abdomen. The three pairs of legs and two pairs of wings arise out of the middle division, the thorax. Insect larvae, however, present a variety of shapes and appendage combinations that strain all definitions of the universal insect. As the identification keys in this guide are based on adult stages, we concentrate here on describing the adult anatomy (see Figure 2). Of the hundreds of named body parts found in insects, only the main parts, as used to identify insects in this guide, are discussed below. The next chapter, which deals with insect life cycles, introduces the main immature body forms.

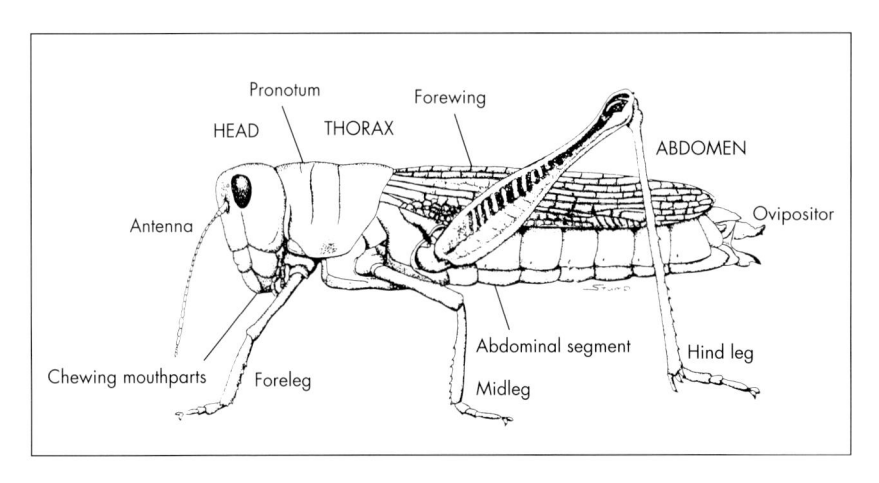

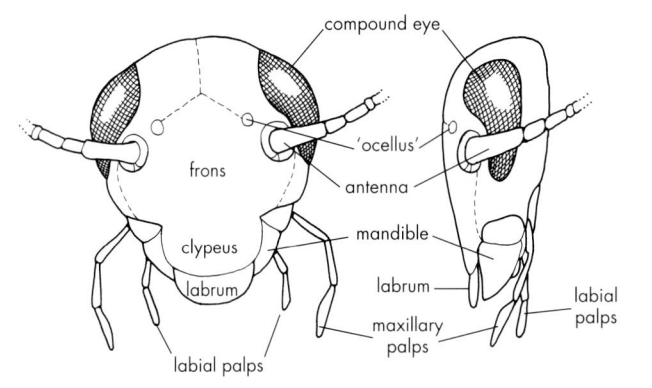

Figure 2. A typical adult insect illustrating the major parts of the body.

Figure 3. The head of a typical insect, a cockroach.

Head

While actually composed of six segments, the insect head appears as a single capsule with mouthparts, two eyes and one pair of antennae. A typical insect with chewing mouthparts is shown, in this case a cockroach (Figure 3). The majority of insect orders follow this pattern, with the movement of the mouthparts from side to side.

The primary mouthparts are the mandibles which form hardened 'jaws' for chewing, sometimes enlarged for catching prey. Behind these are two sets of palps, the maxillary palps and the labial palps which help manipulate the food as well as being organs of taste. A small plate at the front and centre of the head, the labrum, often hides some of these structures from above and is sometimes referred to as an 'upper lip'. This basic set of mouthparts has been adapted to feeding modes other than chewing in some orders of insects (see Figure 4). The bugs (Hemiptera) have most of these structures fused into a tough circular tube called a rostrum which is used for piercing and sucking.

Butterflies and moths (Lepidoptera) have generally lost the mandibles, and the maxillae (which are internal jaw-like structures in chewing insects) are lengthened into a very long coiled proboscis. This uncoils when used for sucking up a liquid diet.

A third major modification is found in the flies (Diptera), where various combinations of parts have fused into a blunt suction pad-like structure which mops up a liquid diet. In some biting flies, like mosquitoes, this is elongated and modified for piercing.

Most insects have one pair of prominent compound eyes. Each eye is made up of a number of separate hexagonal lenses, from just a few in some primitive insects, to many thousands in flying hunters such as dragonflies where the two eyes are so large that they meet in the middle of the head. Some insects also have up to three simple one-lens eyes called ocelli.

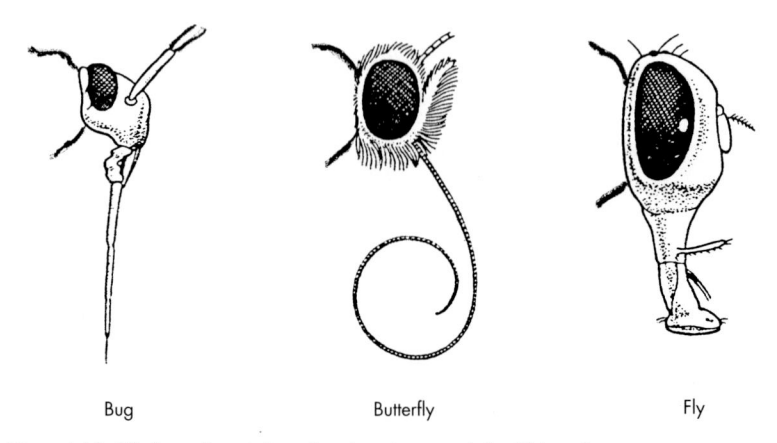

Bug　　　　　　　　　　Butterfly　　　　　　　　　　Fly

Figure 4. Modified mouthparts in various insect groups. [after Chinery]

WHAT IS AN INSECT?

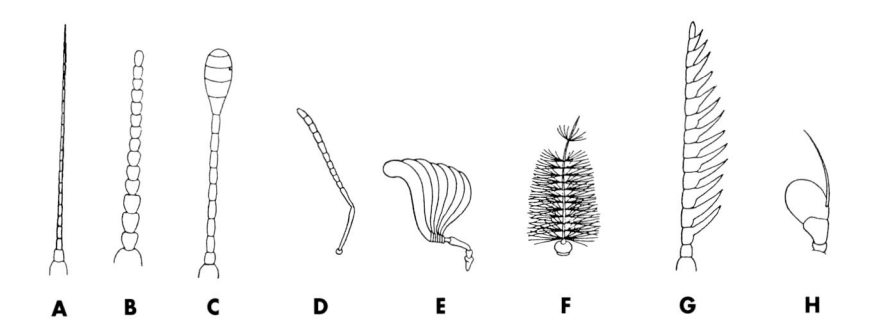

A **B** **C** **D** **E** **F** **G** **H**

Figure 5. The main types of antennae.
A filiform (thread-like) **B** moniliform (bead-like) **C** clubbed (with swollen club-like tip) **D** elbowed (prominently bent, often at 90° and with large scape) **E** lamellate (with last few segments fan-like expansions) **F** plumose (feather-like) **G** serrate (tooth-like or saw-like segments) **H** aristate (short and stout with a thin, bristle-like end)

The last main feature of the head are the antennae, which serve as a major sense organ, including that of smell. The basic antenna is composed of a large basal segment called the scape, and a varying number of segments arising from it. As the shape and number of segments are very important for distinguishing insect groups, Figure 5 illustrates the main types referred to in the text.

Thorax

The middle division of an insect is composed of three subdivisions; viewed from the front these are the prothorax, mesothorax and metathorax. These subdivisions are often not evident. Each subdivision supports one of the three pairs of jointed legs and the last two subdivisions support the wings when present. The first subdivision, as seen from above, is called the pronotum and its shape and size is often used as a diagnostic feature in the text.

Legs

Insect legs come in a great variety of shapes adapted for walking, running, digging, swimming or capturing prey. They are articulated for greater dexterity by having five main movable segments (see Figure 6). The coxa often appears to be a part of the thorax rather than the leg, but is usually movable and is sometimes used as a diagnostic feature. The small trochanter is easy to overlook, but the following three segments are of major importance in insect identification. The femur usually appears as if it is the first segment from the body and is often stockier than the rest. In grasshoppers, for example, it is the largest part of the hind jumping legs. Next is the tibia, which is often the longest segment and is usually thin.

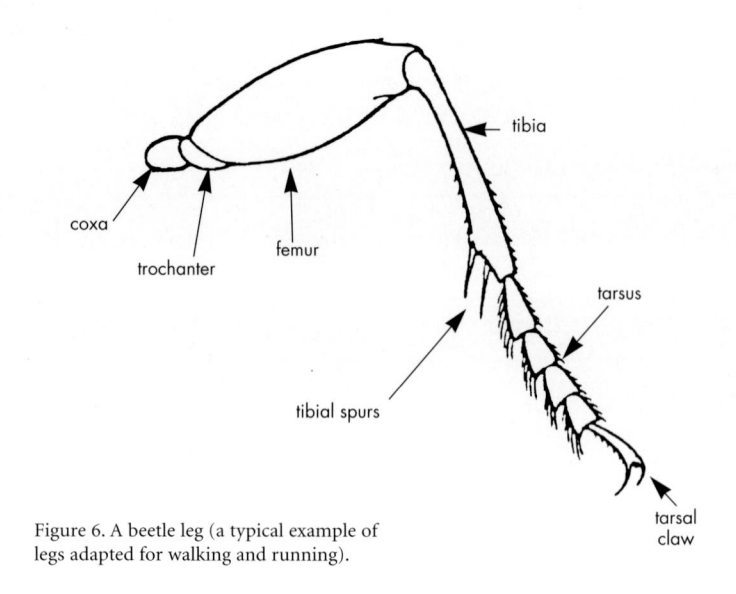

Figure 6. A beetle leg (a typical example of legs adapted for walking and running).

Tibial spurs are prominent at the apex in some insects and are used as a diagnostic feature. The 'feet' of an insect are known as the tarsi, which end with tarsal claws and have from one to five segments.

Wings

Most insects have two pairs of wings as adults, but this is not an absolute rule. Sometimes one or both pairs of wings are reduced in size or even absent in particular species. Often the forewings are modified into toughened covers for the rear wings, as in beetles. Some orders, such as the silverfish (Thysanura) and fleas (Siphonaptera), are always wingless; and others have only one functional pair, such as the flies (Diptera).

These are all diagnostic features, but the most important feature used in insect identification is wing venation. Insect wings have rigid veins which help support the wing in flight. Vein position, size and branching vary between insect groups and are therefore used in identification. Wing veins have names and branch numbers; and even the spaces (cells) between them often have names. Unfortunately, the naming systems have evolved separately for the different insect orders resulting in considerable confusion, with no one system holding true for all insects. Furthermore, different insect orders have different amounts of the full complement of named veins, and vein reduction is common which further confuses the issue. For these reasons, we have decided to avoid this diagnostic feature except in the most obvious cases.

There are several basic wing types mentioned in the text:

- Membranous wings are clear with visible venation, e.g. dragonflies (Odonata). These can be covered in scales, e.g. butterflies (Lepidoptera).
- The toughened forewings of grasshoppers (Orthoptera), which can still beat in flight, are called tegmen.
- The hard wing covers found in beetles (Coleoptera), which do not beat in flight, are called elytra.

Abdomen

The third main body division, the abdomen, contains most of the digestive organs, the breathing apparatus and the reproductive organs. There are 11 segments, although in most orders only 10 are easily discernible.

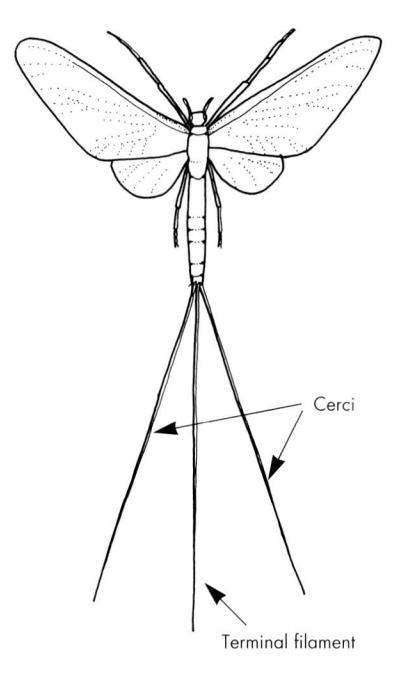

Figure 7. A mayfly showing the types of abdominal appendages.

The eighth and ninth segments contain the reproductive organs, which are mainly concealed internally but can include an external egg-laying device called an ovipositor, which is very long in some wasps and crickets.

The tenth segment sometimes has two thin appendages called cerci. Their presence and length is very diagnostic, as in the long cerci of stoneflies (Plecoptera). In the earwigs (Dermaptera), the cerci are not filamentous but modified into stout pincers; and in the mayflies (Ephemeroptera), there is a third, long, central 'tail' called a terminal filament (Figure 7).

INSECT LIFE CYCLES

There are more than 86 000 species of insects described in Australia, with a large number awaiting names or discovery. With so many separate evolutionary stories, it would be expected that few lifestyle possibilities would be missed by insects. Indeed, there is virtually no organic food source or habitat that is not utilised by insects. All forests, mountains, streams, lakes, deserts and cities in Australia are home to insects living as herbivores, carnivores, parasites and scavengers, with varying degrees of specialisation.

Among the extremes of adaptation are some water striders (Hemiptera) which live out their whole lives on the surface of the open ocean; lice (Phthiraptera) that live on seals which swim in the −2°C Antarctic Ocean; ground beetles (Coleoptera) which live in the totally dark zone of cold caves in Tasmania; aquatic midge larvae (Diptera) which can withstand the drying of waterholes followed by temperatures up to 58°C; and the many insects which not only survive eating very poisonous plants, but store and use the poisons for their own defence in later life.

The different life cycle types found in the insect world are used as diagnostic features at the order level. Insects grow by periodically shedding their entire outer covering, the exoskeleton, which allows the next layer to expand before hardening and make room for growth. Therefore, apparent growth in insects is not smoothly continuous but in stages. Each of these growth stages is called an instar. The number of instars varies among insect orders, averaging between four and eight, with an extreme of 50 in silverfish (Thysanura). Only when an insect is in its last instar, called the imago, is it an adult, capable of reproduction and, in many insects, flight. In all but a few primitive insect orders, growth ceases at the adult stage. Therefore, differences in size evident among individual adults of the same species of insect are mainly due to food availability during the growth of the immature stages.

There are two basic and very different ways in which insects develop to the adult stage. All insects start from an egg; though in some rare cases, such as aphids (Hemiptera) and in some flies, the eggs develop internally, and live young are born. The first instar can then be either a nonreproductive and nonwinged, but recognisable, copy of the adult, or a creature completely different from the adult.

Primitive wingless insects, such as the silverfish, and many of the lower orders of winged insects grow as adult copies which slowly develop their wings on the outside of the body during the later

INSECT LIFE CYCLES

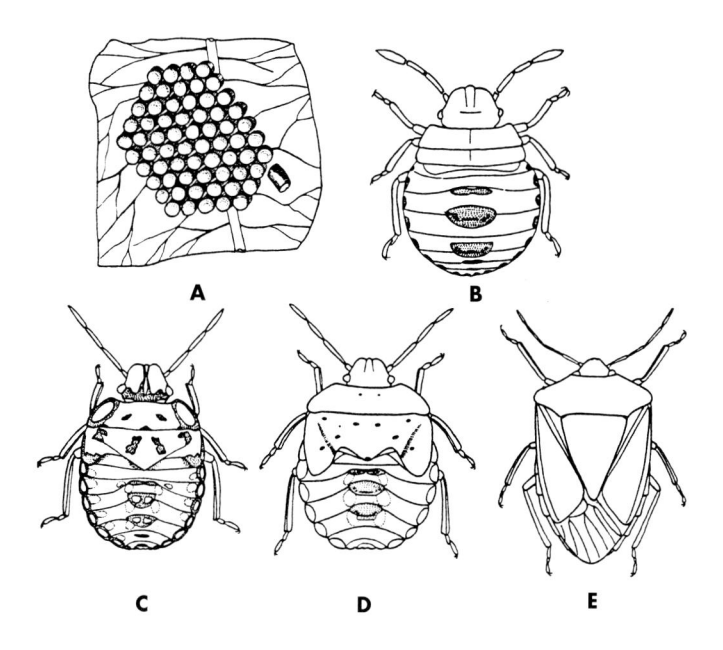

Figure 8. Stages in the incomplete or exopterygote life cycle illustrated by a bug of the family Pentatomidae (order Hemiptera): (a) eggs; (b) first instar nymph; (c) third instar nymph with very early wing buds; (d) final instar nymph with wing sheaths; (e) the adult or imago, with full wings. [after J. Tonnoir]

instars. These are known as exopterygote insects (from *exo* = outside and *pterygote* = wings). This life cycle is also referred to as an incomplete life cycle (Figure 8). These immature stages are usually known as nymphs. Typical examples are grasshoppers, mantids and true bugs, where in the last few instars, wing buds start to develop on the outside and become large enough to be functional in the last moult. These nymphs tend to live the same lifestyle as the adults, with both stages utilising the same food and habitat.

The higher insect orders have developed a different approach which allows them to utilise separate immature and adult food sources and habitats during the one life cycle. The wings develop on the inside of the body, hence the term 'endopterygote' *(endo* = internal and *pterygote* = wings). From the egg hatches a larva, which is usually more or less caterpillar- or grub-shaped. The larvae are often not very mobile creatures but are very efficient food gatherers. The larvae undergo up to 10 instars, growing bigger with each moult but not developing any visible adult characteristics. They then enter a separate, usually stationary stage, during which the body is completely transformed into the adult form. This is the pupal or chrysalis stage, which is characteristic of butterflies, beetles, flies and other orders. This life cycle is referred to as a complete life cycle (see Figure 9).

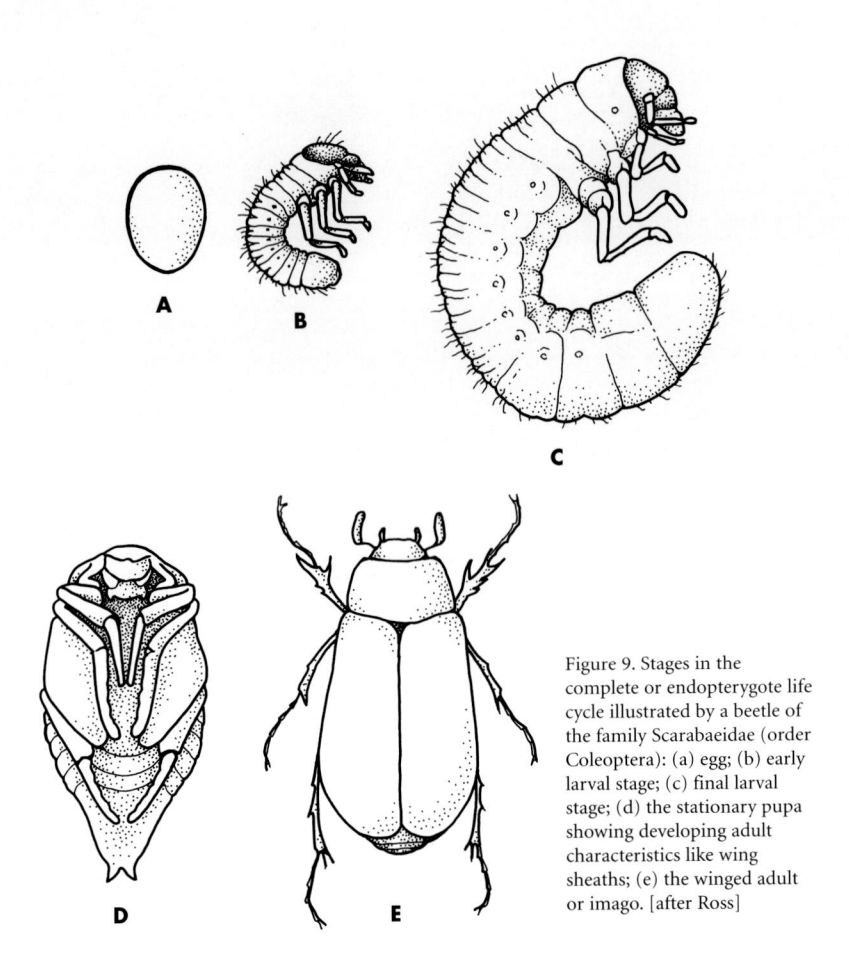

Figure 9. Stages in the complete or endopterygote life cycle illustrated by a beetle of the family Scarabaeidae (order Coleoptera): (a) egg; (b) early larval stage; (c) final larval stage; (d) the stationary pupa showing developing adult characteristics like wing sheaths; (e) the winged adult or imago. [after Ross]

It allows a broad use of habitats. For example, some wasps are parasitic on various insects as larvae and live on a nectar diet as adults; moths and butterflies eat leaves as larvae and the adults fly to flowers to feed on nectar; mosquitoes have detritus-feeding aquatic larvae and the adults fly in search of nectar and blood meals. This life cycle allows the larvae to store so much fat reserve that the adults of some species do not need to spend any time looking for food, but use their time only to reproduce.

CRYPSIS AND MIMICRY

When searching for insects in their natural habitat, one becomes aware that a considerable number of species use various methods of concealment. Insects that have adaptations to help them blend with their environment are termed cryptic.

Many ingenious methods have evolved, based upon the principles of reducing body outline (to avoid casting shadows); disrupting the body shape with bold and confusing patterns; and actually imitating an object like a leaf or twig. Combined, these strategies can result in almost total invisibility. Crypsis is very common and effective. How often have you seen a moth at its daytime roost, despite the fact that over 20 000 species live in Australia?

Ledromorpha sp., **Cicadellidae**. The flange around the flat body of this plant hopper nymph eliminates shadow

The bold, irregular patterns of this desert grasshopper disrupt its body shape, **Acrididae**

ABOVE: This treehopper combines shadow elimination, disruptive patterns and imitation of the bark, **Fulgoridae**

LEFT: *Parasipyloidea* sp., **Phasmatidae**. This tropical stick insect imitates grass in appearance and behaviour (stance)

Often the immature stages of insects use crypsis, while the adults, which can fly and otherwise better evade predators, drop the disguise. Many grasshoppers and moths dabble in crypsis in these early stages.

However, for the purposes of a guide book to insect forms, a different type of disguise needs to be exposed. This is mimicry, the strategy of imitating the form and/or colour patterns of other insect species. Insects which exhibit very strong patterns of colours like red, yellow and black often do so to advertise that they are distasteful or even poisonous to predators. The bold yellow–black pattern of wasps advertises their stinging ability, while the bold stripes of some caterpillars are advertising the poisons they store in their bodies. These genuinely dangerous prey tend to utilise similar warning patterns despite not being otherwise related—a phenomenon called 'convergent evolution'. This has also allowed the processes of natural selection to favour the imitation of these patterns and shapes by other insects that are quite edible. These are 'Batesian mimics', and besides fooling predators, they can also surprise the field taxonomist. A very common form to mimic is the generalised shape and colour of wasps. There are whole chains of such mimics across several orders of insects. Many flies in the family Stratiomyidae mimic the typical stinging wasp form, while a fly in the family Therevidae mimics the fearsome spider-hunting wasps of the family Pompilidae. Braconid wasps are mimicked by moths of the family Oecophoridae and by sap-sucking bugs of the family Miridae.

The trick works in nature as long as there are more models than mimics. Otherwise predators will not learn the lesson of dangerous appearance through experience often enough.

Fly, *Syndipnomyia* sp., **Stratiomyidae**, mimics stinging wasp

Fly, *Agapophytus* sp., **Therevidae**, mimics hunting wasp (family Pompilidae)

Model for mimics: wasp, **Braconidae**

Moth, *Pseudaegeria* sp., **Oecophoridae**, mimics braconid wasps

Model for mimics: spider wasp, **Pompilidae**

Model for mimics: beetle, *Metriorrhynchus* sp., **Lycidae**

Bug, **Miridae**, mimics braconid wasp

A FIELD GUIDE TO INSECTS

Moth, Cyaria sp., **Arctiidae** mimics *Metriorrhynchus*

Bug nymph, *Riptortus* sp., **Alydidae**, mimics ants

Katydid nymph, **Tettigoniidae**, mimics ants

The longest chain of Batesian mimicry in Australia involves several orders and tens of families. The model here is a brick-red coloured genus of beetles in the family Lycidae (*Metriorrhynchus*), which has many widely distributed species to keep the balance of the deception functioning. One of its many possible mimics is a moth in the family Arctiidae.

Ants have few predators due to their pugnacious and often chemically unpleasant nature and so tend to be a common model. Many insects imitate a generalised ant shape for the first one or two instars of their precarious development. Stick insects, bugs, crickets and mantids often employ this defence. The other surprises in this chain are the many spiders that mimic ants. These can be true masters, with a proper 'waist' like ants, the same frantic walking style and with a habit of holding their superfluous first pair of legs aloft to imitate antennae.

To a human observer, these deceptions work well on a superficial level; but once you have one of these mimics in the hand, the true characteristics of the order to which they really belong can be discerned. Thus suspicious looking wasps turn out to be flies because they still have only one pair of wings like true flies do; suspicious looking ants and wasps can turn out to be bugs because in profile their beak-like mouthparts still stand out instead of the expected biting jaws; and the spiders will sometimes drop their guard and use their 'antennae' to walk on, and their set of eight spider eyes is evident in close up.

Jumping spider, **Salticidae**, mimics ants

While many mimics adhere to a generalised body plan, there are also specialists which imitate individual species and these can be very difficult to spot. When using this guide to gain a general understanding of insect identification, do not take appearances for granted. Finding and observing the crypsis and mimicry masters in the field is a very rewarding challenge.

Further Reading

COTT, H.B. (1940), *Adaptive Coloration in Animals*, Methuen, London.
EDMUNDS, M. (1981), 'On defining mimicry', *Biological Journal of the Linnean Society* 16, pp. 9–10.
OWEN, D.F. (1980), *Camouflage and Mimicry*, Oxford University Press, Oxford.
Pasteur, G. (1982), 'A calssificatory review of mimicry systems', *Annual Review of Ecological Syst.* 13, pp. 169–99.
WICKLER, W. (1968), *Mimicry in Animals and Plants*, Weidenfeld and Nicolson, London.
ZBOROWSKI, P. (1991), *Animals in Disguise*, Currawong Press/Reed, Sydney.

COLLECTING INSECTS

The following characteristics of insects make their study different from that of vertebrates:

- Small size: the majority of species are less than 5 mm long.
- Diversity: insects are the most diverse group of (visible) organisms on the planet, often with many closely related species with overlapping distributions.
- Large numbers of individuals: populations of thousands of individuals often occur in small areas and even 'rare' species are usually so only because of restricted habitats or camouflage.

As a result, entomologists since the time of Linnaeus in the 18th century have found it necessary to maintain collections of preserved insect specimens to enable identification by comparison of fine characteristics in a series of individuals. Outside the large national insect collections in museums, the work of amateur collectors has also played an important role in the study of the Australian fauna.

While most insect species are too abundant to have numbers affected by the activities of collectors, there is an increasing number of species which could become vulnerable and these are protected by law in different States. Most National Park systems restrict collecting without permits. We thus advise insect collectors to learn of any such regulations in their area before collecting.

Collecting Methods

Many techniques for systematic sampling of insects in ecological studies have been developed. Here we discuss the simpler collecting methods, keeping in mind that patience, a good eye and experience are central to them all.

Hand Collecting

Many medium to large insects can be collected directly off foliage, blossom, tree trunks, logs or the ground. Looking under logs or rocks is also very useful because many nocturnal insects hide in dark recesses during the day. Captured specimens can then be placed directly into a killing jar or alcohol jar (with 70% ethyl alcohol) and mounted later. A typical killing jar for most insects other than butterflies and moths has a plaster-of-paris base to absorb a sprinkle of ethyl acetate.

Insect Nets

There are many types of nets available. A typical net used for flying insects is an aluminium hoop about 40–50 cm in diameter on an extendable handle, with a net bag of fine muslin cloth with a canvas edge. Small wasps, flies and some other insect groups are better caught by a sweep net. In these, the net is enclosed by a tough canvas outer bag which protects the net fabric while it is swept through bushes and grasses. Some have chicken-wire placed over the net opening to exclude branches and leaves. For aquatic insects a dip net is used. This net has a triangular frame so that one flat edge can be swept along the bottom where many insects sit.

Pitfall Traps

Many ground-living insects are only active at night and are best caught using pitfall traps that are set in a particular area for some time. These traps consist of a strong container about 10 cm in diameter and depth that is buried up to its rim. Preservative (a 70:30 mix of water and car antifreeze) is added to a depth of 3 cm. A funnel (with a 2-cm lower opening) placed across the top can prevent larger animals like lizards from falling in as well as slow down evaporation of the preservative. A strong roof is placed about 3 cm aboveground to keep out rain and larger curious vertebrates. Pitfalls should be emptied by sieving the contents at regular intervals of several days to a few weeks. This technique is often used for long-term seasonal studies of diversity.

Light

Many insects are attracted to lights, especially on hot, humid, windless nights. The simplest method to use this behaviour for collecting is to suspend a light (mains, battery or generator powered) in front of a hanging white sheet and pick off the insects of interest. Mercury vapour and ultraviolet fluorescent lights are more attractive to insects than are standard light globes. The type and number of insects attracted to lights vary between seasons, moon phases and local weather, so it is worth sampling a site many times.

Mounting and Preserving

Museum insect collections have been maintained for over 200 years, and correctly preserved and stored specimens are still in a condition to be of great value to taxonomy. Most insect specimens are either pinned or stored in 70% ethyl alcohol, often with additives to reduce colour loss or hardening in particular groups.

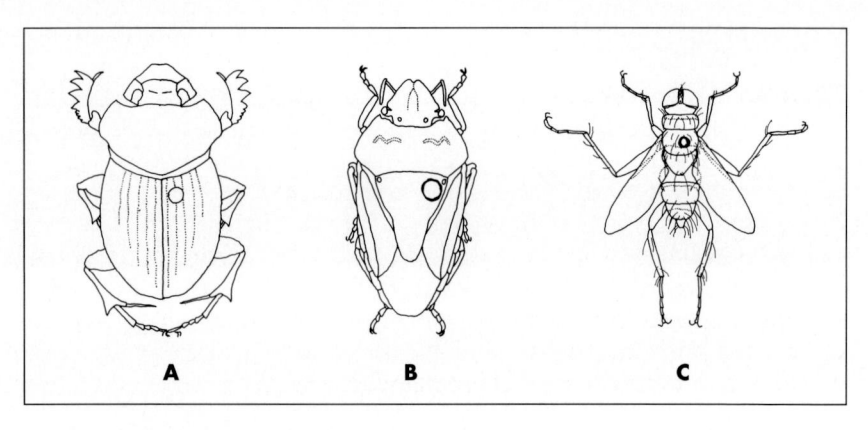

Figure 10. Standard pin positions for: (a) beetles (Coleoptera); (b) bugs (Hemiptera); (c) flies (Diptera) and wasps (Hymenoptera).

Some small, soft-bodied insects such as aphids and scales are best stained and mounted on microscope slides. It is important that all insects are mounted in such a way that the important diagnostic features for that group (e.g. particular appendages or the genitalia) are exposed or accessible for further study.

The majority of insects are pinned before storage in specially made boxes or trays. A pinning box is lined with cork or fine plastic foam, and needs a tight-fitting lid to keep out scavenging insect pests which can eat out the collection. A muslin bag with naphthalene, pinned into the corner of the box, will also guard against these pests, and a cool dry place for storage will minimise problems with fungal damage.

Special long, stainless steel insect pins are used to pin insects. These are available in various sizes. Collectors are advised not to use any pins thinner than No. 3 pins for smaller insects, or thicker than No. 5 for larger ones. Pin specimens as soon as possible to avoid brittleness. The correct pinning position varies between orders, but is usually through the thorax (or elytra in beetles), with the specimen placed in the top third of the pin, leaving room below for labels (Figure 10).

A safe way to pin an insect is onto a sheet of 3 cm thick plastic foam, bracing and setting the appendages with other pins. Some insects such as butterflies and moths need to have their wings displayed and a setting board is used for this task (Figure 11). The abdomen fits into a central groove and the wings are spread and held in position on the cork sides by pinned strips of tracing paper. The wings should be set with the lower margins of the forewings and the upper margins of the hind wings about 90° to the body, making special effort not to remove scales.

Insects smaller than 1 cm can be set using various types of double mounts. Short micropins are available, but a safer and more popular

COLLECTING INSECTS

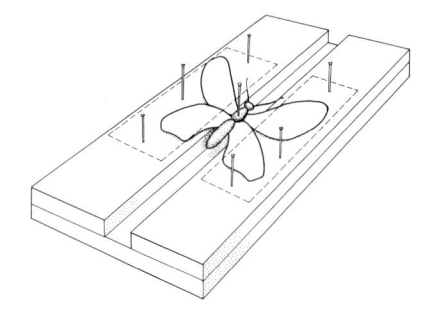

Figure 11. An insect setting board.

Figure 12. An insect mounted onto a card triangle ready for pinning into a storage box.

alternative is to glue the specimen to the tip of a small sturdy card triangle. This mount is then pinned into the storage box by a No. 5 pin (Figure 12). A minimal amount of clear-drying, water-soluble glue is used and it is important that diagnostic features are left visible.

The last essential stage is labelling. Every specimen should have its own label, about 8 mm by 16 mm, placed midway on the pin, or inside the alcohol jar with wet specimens. An indelible ink pen such as a 0.25 Rapidograph or similar is best. An example of the standard minimum information required is:

Location (with State)	10 km NW of Mareeba, Qld
Date (with month in roman numerals)	19. IX. 1994
Name of collector	R.I. Storey

Extra information such as collection method, host plant or altitude may be added on a second label if required. A rough pencil label should be used at the time of capture, as errors in the permanent label take away the specimen's value.

The above outline is very introductory. It is recommended that interested readers obtain one of the several inexpensive small books on the subject such as *Methods For Collecting, Preserving and Studying Insects and Allied Forms* by Murray S. Upton, published by the Australian Entomological Society, 1991.

Several State museums have similar publications. Much of the equipment for catching, preserving and storage of insects, as well as books, can be obtained via a catalogue from:

Australian Entomological Supplies
PO Box 250 Bangalow NSW 2479
http://www.entosupplies.com.au/Frames/menu.html

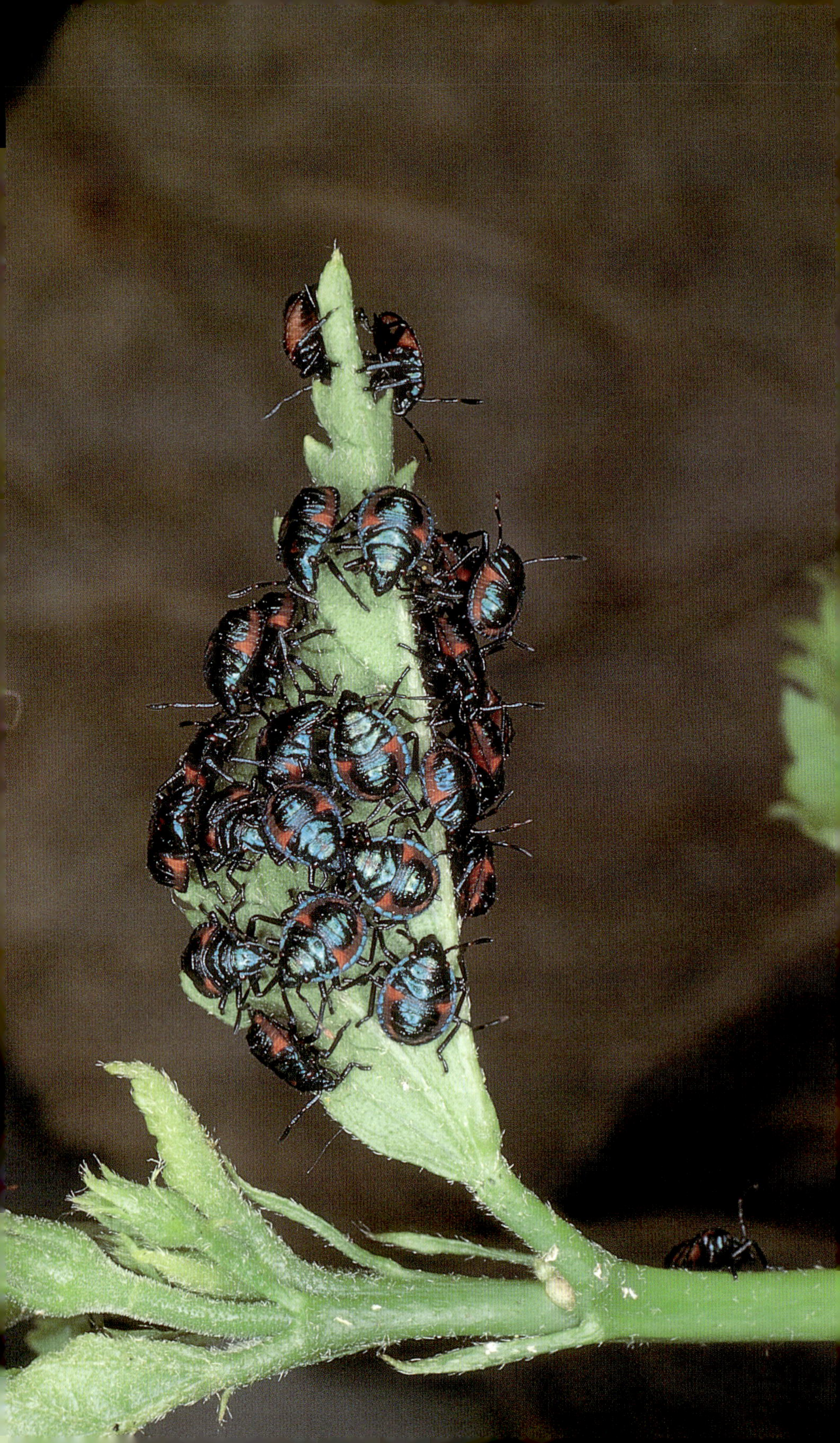

CLASSIFICATION AND A KEY TO THE INSECT ORDERS

Classification

Insects compose the largest class of organisms, the class Insecta, and are placed within the phylum Arthropoda. Classes are divided into orders; and class Insecta is divided into 29 orders, of which 26 are present in Australia. For example, all beetles are in the order Coleoptera and all flies are in the order Diptera. With a few exceptions, order names end with the letters *ra*.

Each order is further divided into families, e.g. within the beetle order, weevils and scarabs are separate families. There are 661 families of insects in Australia alone and closer to 1000 worldwide. In the smallest order, Archaeognatha, there is only one family, while in the largest order, Coleoptera (the beetles), there are over a hundred. All family names end with the letters *idae*.

Orders with many families often group similar families into superfamilies. The names of the superfamilies always end with the letters *oidea*. Some larger families are split into subfamilies and these always end with the letters *inae*.

Each family contains from one to several thousand species. Closely related species are grouped within genera. A genus can have from one to hundreds of species, and these species generally share a similar recognisable form. A taxonomist will often recognise a genus in the field, though it may take detailed laboratory analysis to determine the exact species.

Once an insect is identified to species level it is always cited as genus and species. This format is called the binomial system and was first popularised by Linnaeus, in 1758. In this system, the generic (genus) name is always capitalised; the specific (species) name is not; both names are written in *italics*. Sometimes there is a person's name listed after the species name. This is the name of the taxonomist who first described and named that particular species.

The actual names are derived mainly from Latin, sometimes from Greek and more rarely from other languages; they are, however, always Latinised. These names are often descriptive of the organism in question. For example, the fly order Diptera is derived from *di* = two and *pteron* = wing, as all true flies have only one pair of wings; the butterfly and moth order Lepidoptera is derived from *lepis* = scales and *pteron* = wings, due to the dense scales on all species.

LEFT: 2nd stage nymphs of cotton harlequin bug, *Tectocoris diophthalmus*, Scutelleridae (5 mm long)

In faunal works, the 29 orders of insects are arranged along evolutionary lines, usually listed roughly from the most primitive to the most advanced. This separates the orders into groups with similar life cycles. Following is a list of the Australian insect orders.

Australian Insect Orders

The Pseudo Insects — The Entognatha

These are three classes of primitive, wingless insects which are now placed outside the class Insecta.

Class Collembola	springtails
Class Protura	proturans, minute soil dwellers
Class Diplura	diplurans, larger soil dwellers

Class Insecta

PRIMITIVE, WINGLESS INSECTS — THE APTERYGOTA

Order Archaeognatha	bristletails
Order Thysanura	silverfish

WINGED INSECTS (ADULT STAGE) — THE PTERYGOTA

(a) **Exopterygota**

Incomplete life cycle— insects with immature stages (nymphs) that are similar to the adults and develop their wings on the outside.

Order Ephemeroptera	mayflies
Order Odonata	dragonflies and damselflies
Order Plecoptera	stoneflies
Order Blattodea	cockroaches
Order Isoptera	termites
Order Mantodea	mantids
Order Dermaptera	earwigs
Order Orthoptera	grasshoppers and crickets
Order Phasmatodea	stick and leaf insects
Order Embioptera	web spinners or embiids
Order Psocoptera	including booklice
Order Phthiraptera	lice
Order Hemiptera	including true bugs, hoppers, aphids and scale insects
Order Thysanoptera	thrips

(b) **Endopterygota**

Complete life cycle — insects with immature stages (larvae) different from the adults and that undergo a pupa stage where wings develop on the inside.

Order Megaloptera — alderflies and dobsonflies
Order Neuroptera — including lacewings and antlions
Order Coleoptera — beetles
Order Strepsiptera — stylopids (small insect parasites)
Order Mecoptera — scorpion flies
Order Siphonaptera — fleas
Order Diptera — flies
Order Trichoptera — caddis flies
Order Lepidoptera — moths and butterflies
Order Hymenoptera — sawflies, wasps, bees and ants

Taxonomic Key

Groups of organisms, including insects, are usually identified by using a taxonomic key. A key is basically a series of questions, usually grouped into couplets, which ask about the presence or absence of visible or biological characteristics. Each specimen is eventually

Tiger beetles mating, *Megacephala* sp., **Carabidae–Cicindelinae** (22 mm long)

placed in a named category by a process of elimination. Keys vary in complexity according to the number of species in the group being keyed and the level of accuracy required. Keys to orders are easier to use than keys to species because the characteristics used to separate orders are more basic and therefore easier to observe and describe.

For this guide we have provided a key to the level of order, using simple language where possible. It consists of 30 couplets and has drawings of a typical species for each order. It must be stressed, however, that a key such as this, designed for ease of use, will not identify every insect examined. Immature stages are not included as the key would become too lengthy and would require precise technical language and the reader would often require a microscope. In addition, as in much of biology, for every rule there are exceptions. For example, ants and termites are grouped under winged insects because their reproductive castes are winged, even though the reader would be most likely to collect a wingless worker caste. Similarly, scale insects would be hard to identify as belonging to the order of (often winged) bugs Hemiptera, because they are wingless and have very reduced segmentation and appendages. The most recent key designed to identify all stages of all insects to order is that of Lawrence (1991). This key has 116 couplets and highly technical language and is thus unsuitable for a reader looking for an introduction to the world of insects.

The key presented here, combined with more detailed descriptions and illustrations of each order in the following chapters, is suitable to identify the vast majority of adult insects in the hand.

KEY TO THE INSECT ORDERS

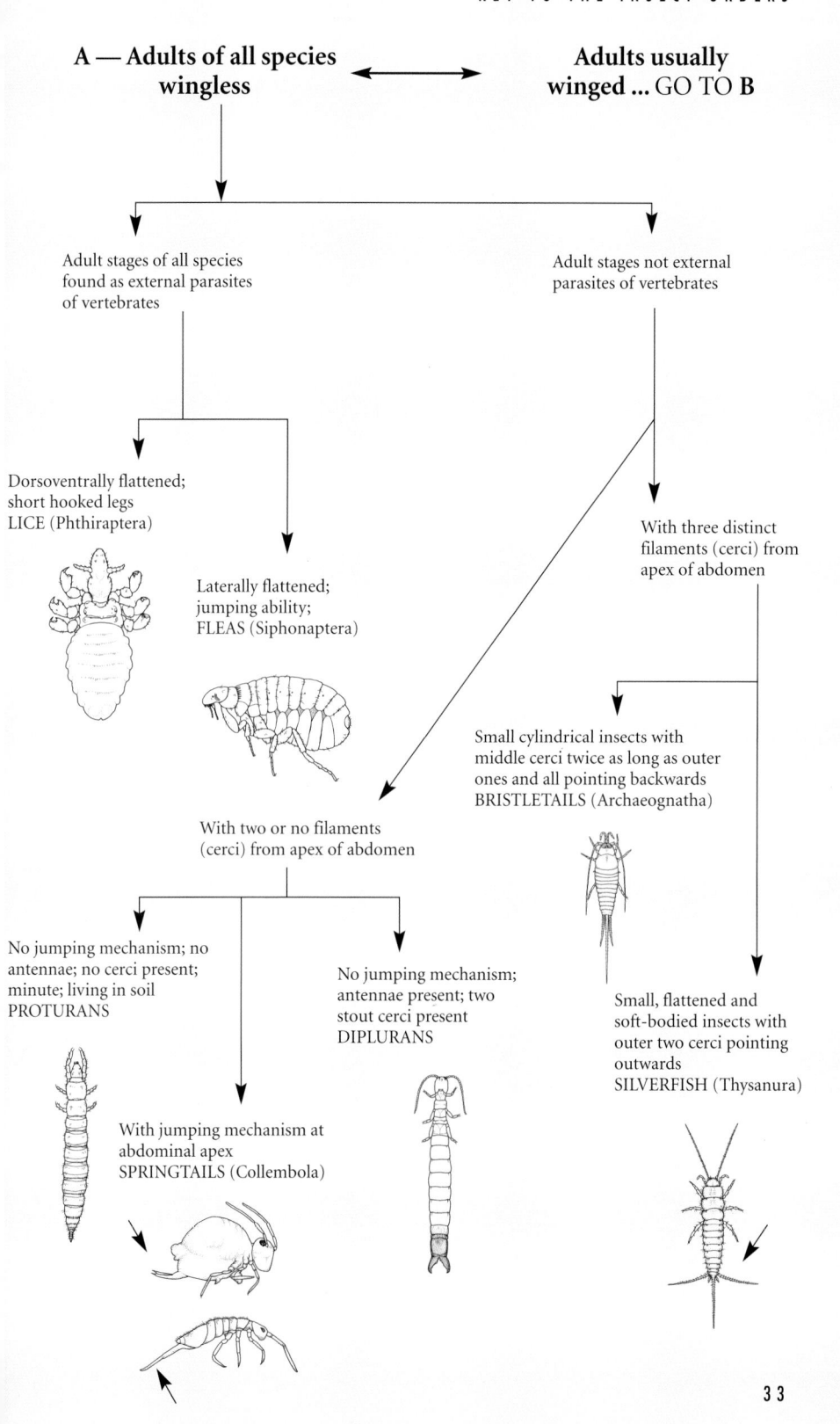

A — Adults of all species wingless ⟷ **Adults usually winged ... GO TO B**

Adult stages of all species found as external parasites of vertebrates

Adult stages not external parasites of vertebrates

Dorsoventrally flattened; short hooked legs
LICE (Phthiraptera)

Laterally flattened; jumping ability;
FLEAS (Siphonaptera)

With three distinct filaments (cerci) from apex of abdomen

Small cylindrical insects with middle cerci twice as long as outer ones and all pointing backwards
BRISTLETAILS (Archaeognatha)

With two or no filaments (cerci) from apex of abdomen

No jumping mechanism; no antennae; no cerci present; minute; living in soil
PROTURANS

No jumping mechanism; antennae present; two stout cerci present
DIPLURANS

Small, flattened and soft-bodied insects with outer two cerci pointing outwards
SILVERFISH (Thysanura)

With jumping mechanism at abdominal apex
SPRINGTAILS (Collembola)

A FIELD GUIDE TO INSECTS

B—Adults usually winged

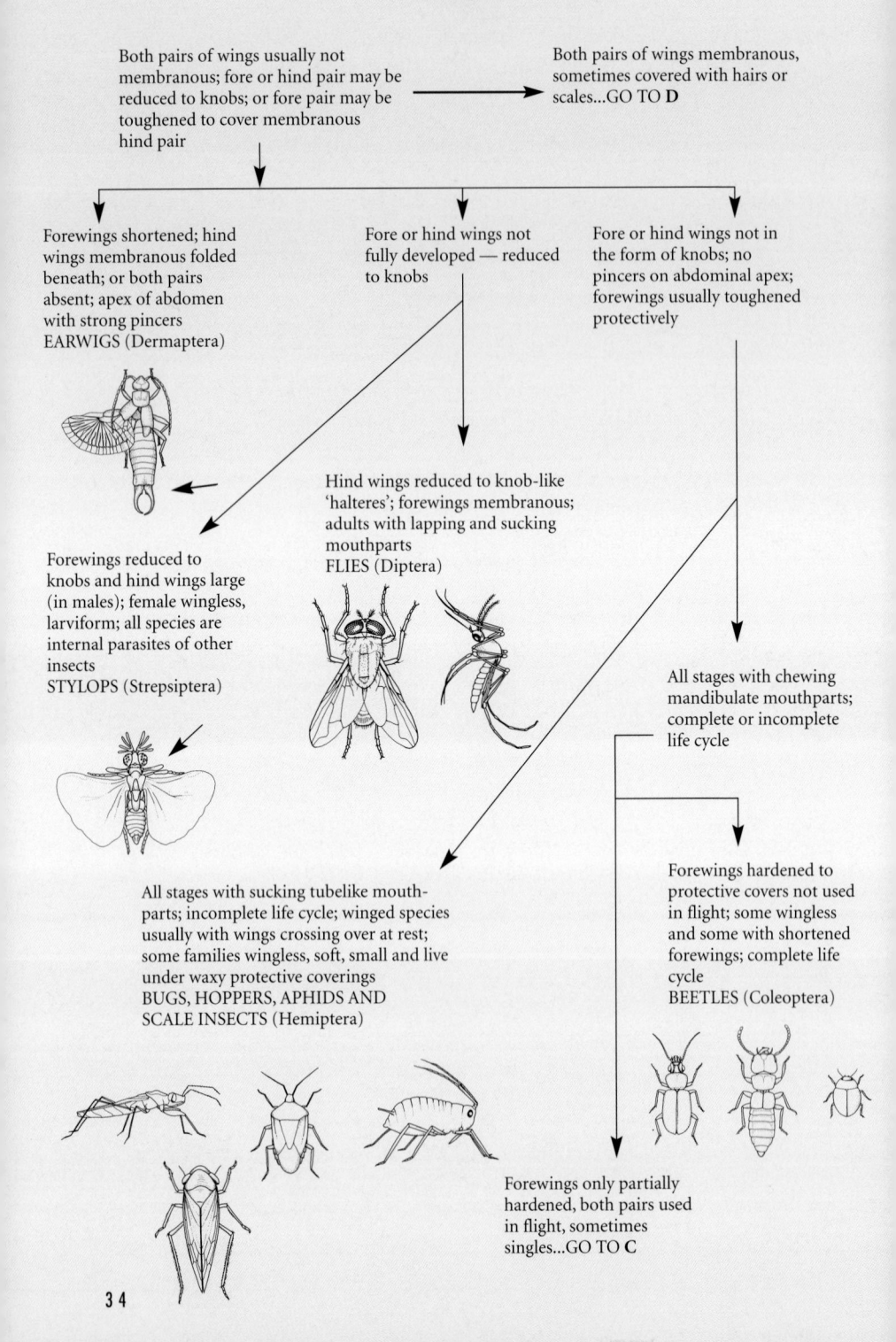

Both pairs of wings usually not membranous; fore or hind pair may be reduced to knobs; or fore pair may be toughened to cover membranous hind pair → Both pairs of wings membranous, sometimes covered with hairs or scales...GO TO **D**

Forewings shortened; hind wings membranous folded beneath; or both pairs absent; apex of abdomen with strong pincers
EARWIGS (Dermaptera)

Fore or hind wings not fully developed — reduced to knobs

Fore or hind wings not in the form of knobs; no pincers on abdominal apex; forewings usually toughened protectively

Forewings reduced to knobs and hind wings large (in males); female wingless, larviform; all species are internal parasites of other insects
STYLOPS (Strepsiptera)

Hind wings reduced to knob-like 'halteres'; forewings membranous; adults with lapping and sucking mouthparts
FLIES (Diptera)

All stages with chewing mandibulate mouthparts; complete or incomplete life cycle

All stages with sucking tubelike mouthparts; incomplete life cycle; winged species usually with wings crossing over at rest; some families wingless, soft, small and live under waxy protective coverings
BUGS, HOPPERS, APHIDS AND SCALE INSECTS (Hemiptera)

Forewings hardened to protective covers not used in flight; some wingless and some with shortened forewings; complete life cycle
BEETLES (Coleoptera)

Forewings only partially hardened, both pairs used in flight, sometimes singles...GO TO **C**

34

KEY TO THE INSECT ORDERS

C — Forewings only partially hardened, both pairs used in flight; sometimes wingless

Usually enlarged jumping hind legs; often with a file-like mechanism to produce 'song'
CRICKETS AND GRASSHOPPERS (Orthoptera)

Hind legs not enlarged for jumping; do not 'sing'

Stick or leafshaped; usually very long and thin; sometimes very large; herbivores
STICK AND LEAF INSECTS (Phasmatodea)

Not stick or leafshaped; no raptorial front legs or very large eyes; usually flattened; active scavenger
COCKROACHES (Blattodea)

Raptorial front legs; large eyes in a triangular and very mobile head; predators
MANTIDS (Mantodea)

D — Both pairs of wings membranous; sometimes covered with hairs or scales

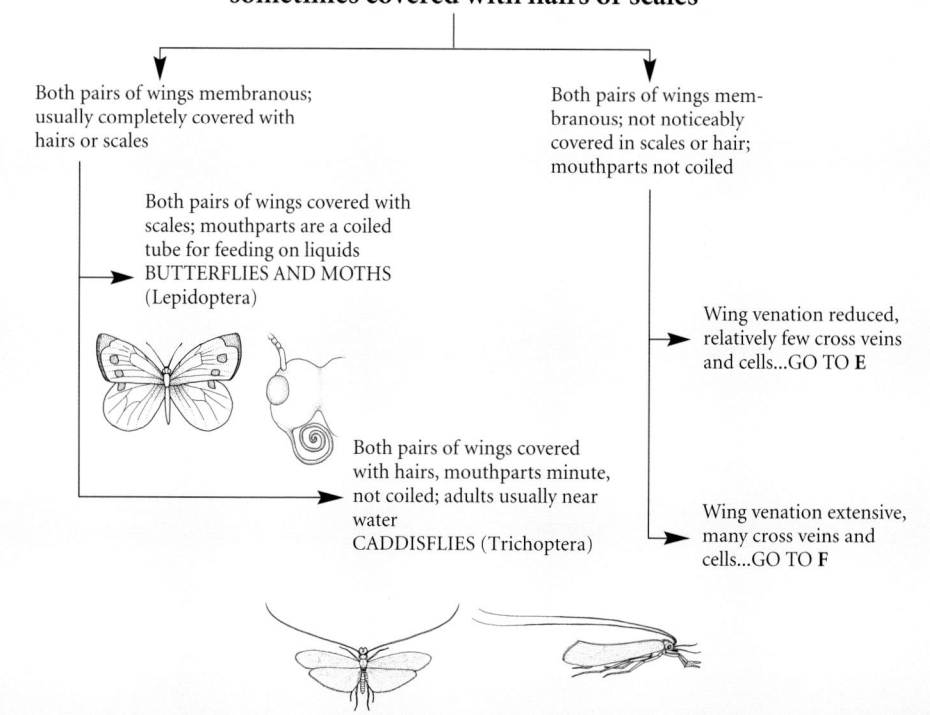

Both pairs of wings membranous; usually completely covered with hairs or scales

Both pairs of wings membranous; not noticeably covered in scales or hair; mouthparts not coiled

Both pairs of wings covered with scales; mouthparts are a coiled tube for feeding on liquids
BUTTERFLIES AND MOTHS (Lepidoptera)

Both pairs of wings covered with hairs, mouthparts minute, not coiled; adults usually near water
CADDISFLIES (Trichoptera)

Wing venation reduced, relatively few cross veins and cells...GO TO **E**

Wing venation extensive, many cross veins and cells...GO TO **F**

35

A FIELD GUIDE TO INSECTS

E — Wing venation reduced, relatively few cross veins and cells

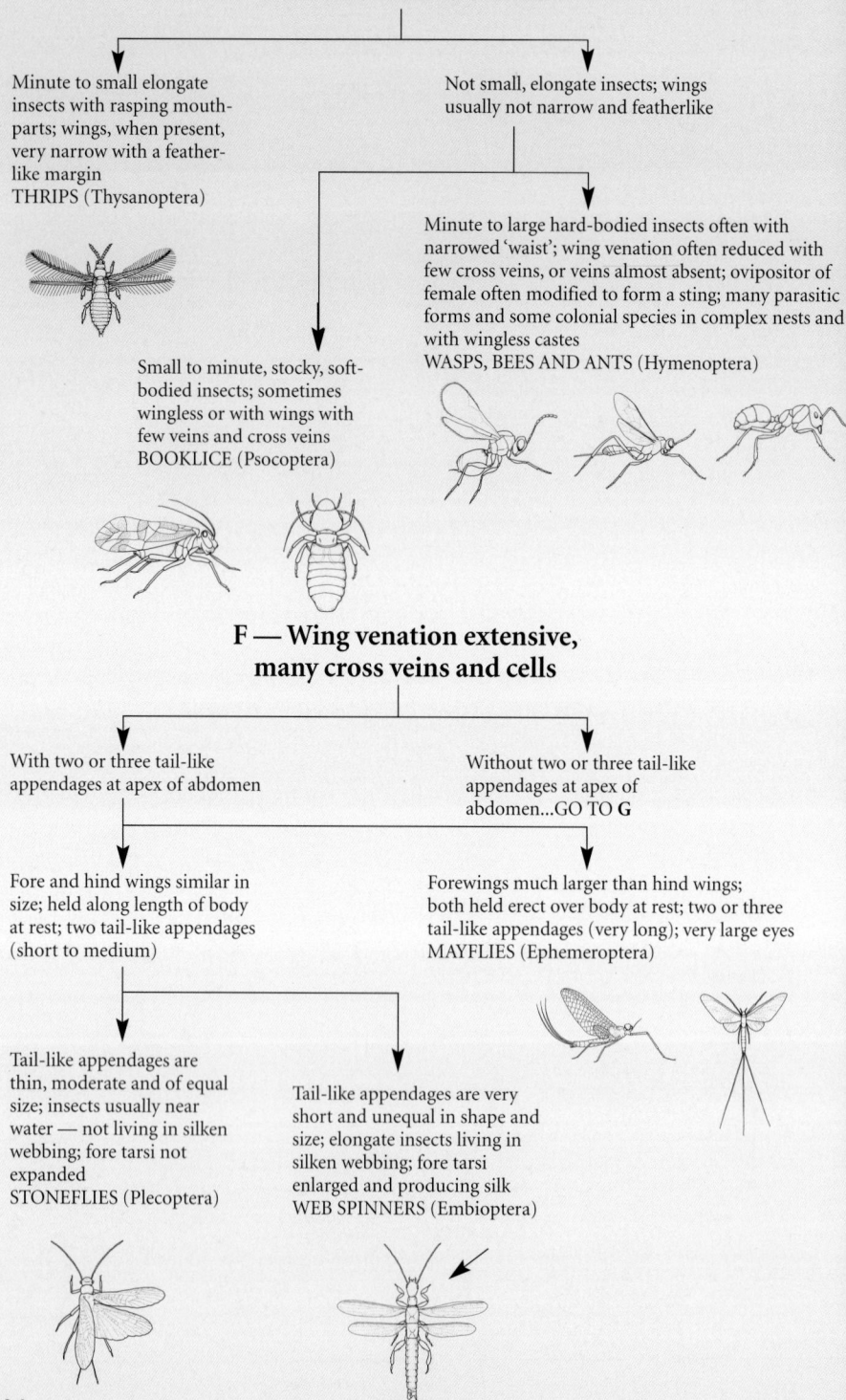

Minute to small elongate insects with rasping mouthparts; wings, when present, very narrow with a featherlike margin
THRIPS (Thysanoptera)

Not small, elongate insects; wings usually not narrow and featherlike

Small to minute, stocky, soft-bodied insects; sometimes wingless or with wings with few veins and cross veins
BOOKLICE (Psocoptera)

Minute to large hard-bodied insects often with narrowed 'waist'; wing venation often reduced with few cross veins, or veins almost absent; ovipositor of female often modified to form a sting; many parasitic forms and some colonial species in complex nests and with wingless castes
WASPS, BEES AND ANTS (Hymenoptera)

F — Wing venation extensive, many cross veins and cells

With two or three tail-like appendages at apex of abdomen

Without two or three tail-like appendages at apex of abdomen...GO TO **G**

Fore and hind wings similar in size; held along length of body at rest; two tail-like appendages (short to medium)

Forewings much larger than hind wings; both held erect over body at rest; two or three tail-like appendages (very long); very large eyes
MAYFLIES (Ephemeroptera)

Tail-like appendages are thin, moderate and of equal size; insects usually near water — not living in silken webbing; fore tarsi not expanded
STONEFLIES (Plecoptera)

Tail-like appendages are very short and unequal in shape and size; elongate insects living in silken webbing; fore tarsi enlarged and producing silk
WEB SPINNERS (Embioptera)

KEY TO THE INSECT ORDERS

G — Without two or three tail-like appendages at apex of abdomen

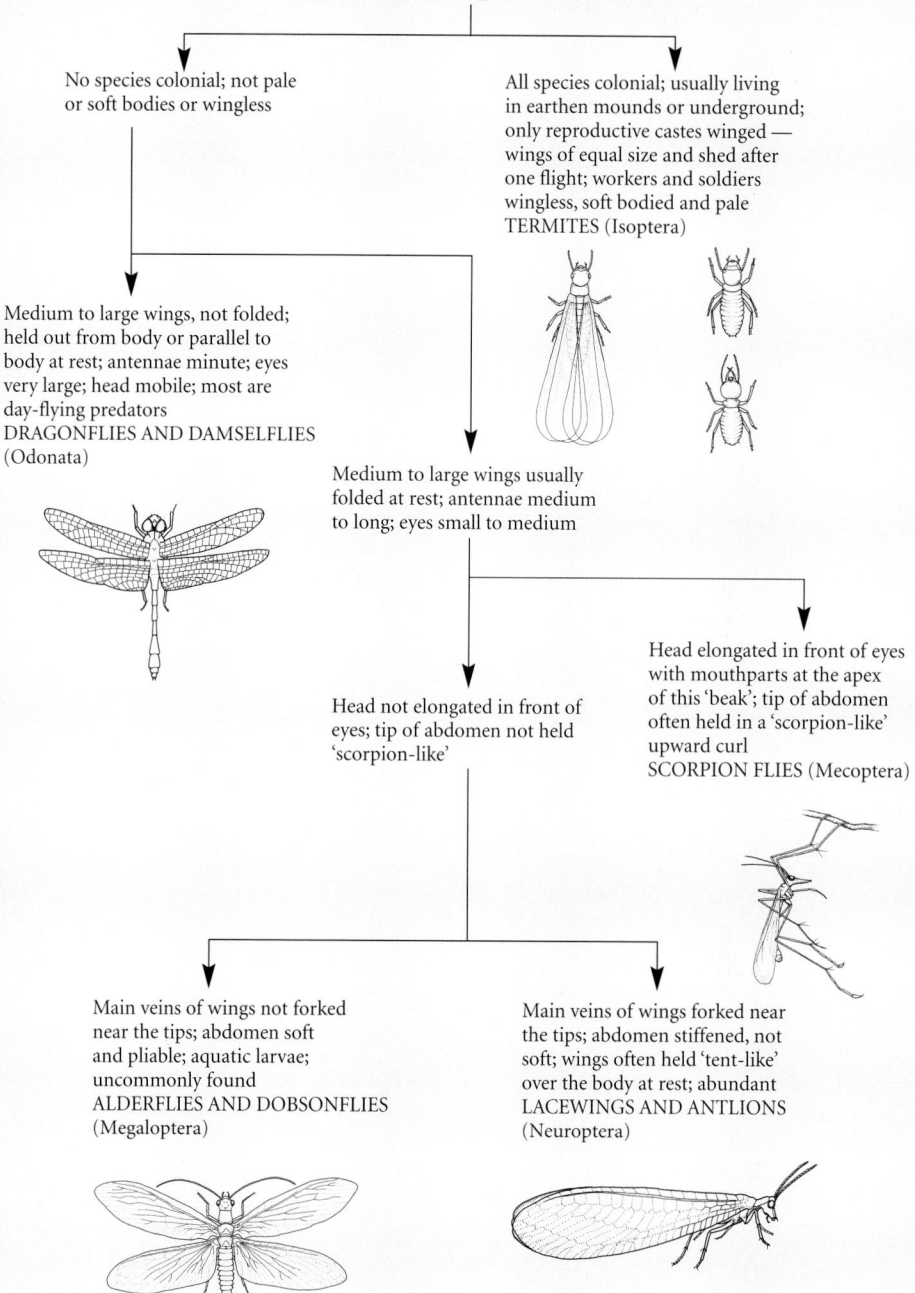

No species colonial; not pale or soft bodies or wingless

All species colonial; usually living in earthen mounds or underground; only reproductive castes winged — wings of equal size and shed after one flight; workers and soldiers wingless, soft bodied and pale
TERMITES (Isoptera)

Medium to large wings, not folded; held out from body or parallel to body at rest; antennae minute; eyes very large; head mobile; most are day-flying predators
DRAGONFLIES AND DAMSELFLIES (Odonata)

Medium to large wings usually folded at rest; antennae medium to long; eyes small to medium

Head not elongated in front of eyes; tip of abdomen not held 'scorpion-like'

Head elongated in front of eyes with mouthparts at the apex of this 'beak'; tip of abdomen often held in a 'scorpion-like' upward curl
SCORPION FLIES (Mecoptera)

Main veins of wings not forked near the tips; abdomen soft and pliable; aquatic larvae; uncommonly found
ALDERFLIES AND DOBSONFLIES (Megaloptera)

Main veins of wings forked near the tips; abdomen stiffened, not soft; wings often held 'tent-like' over the body at rest; abundant
LACEWINGS AND ANTLIONS (Neuroptera)

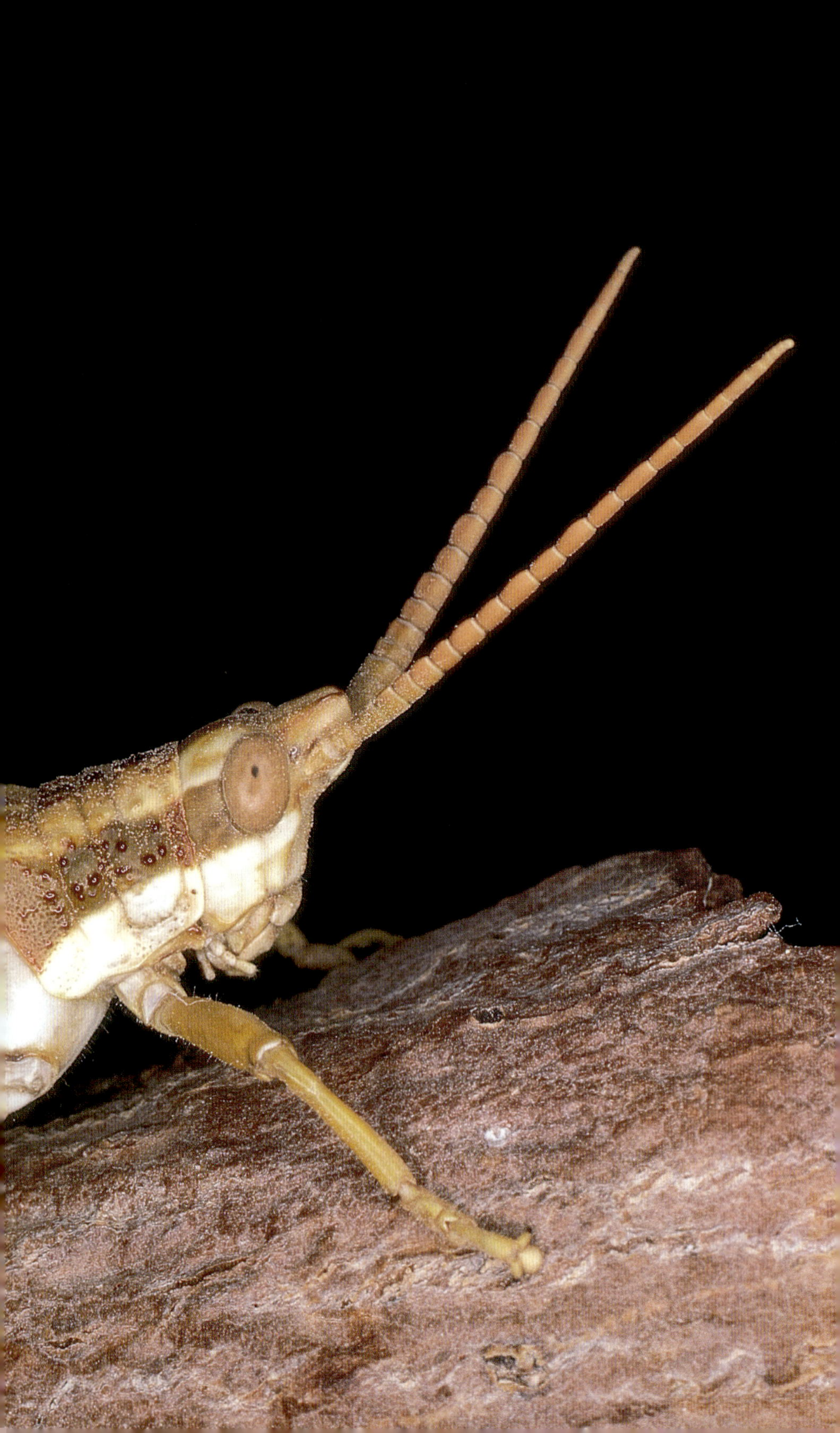

SPRINGTAILS, PROTURANS, DIPLURANS
THE ENTOGNATHA

Springtails (Class Collembola)
World families: 20; species: 6000 **Australian families:** 14; species: 1630

Proturans (Class Protura)
World families: 4; species: 500 **Australian families:** 3; species: 30

Diplurans (Class Diplura)
World families: 9; species: 800 **Australian families:** 5; species: 31

The taxonomic status of these groups has varied over the years. All these primitive insects have the required six legs and roughly three body divisions, but modern classifications place them outside the class Insecta. They are known collectively as 'entognathous hexa-pods'. Entognathy is a state in which the mouthparts, like the mandibles, are partly obscured by oral folds or cheeks, and are therefore more or less internal. Hexapoda simply means having six legs and is a term applied to the true insects as well.

These three divisions were once orders within the class Insecta, but are now considered as three separate classes.

Characteristics of Classes

The following descriptions emphasise how the three classes differ from each other and from the superficially similar true insects.

Springtails (Class Collembola)
Small (usually 1–3 mm, rarely up to 10 mm in body length); body of two main shapes: compact and almost spherical, or elongate and cylindrical; wingless; many have a forked 'tail' or anal spine which can flex to propel the body into a high jump; head with thick four-segmented, moniliform antennae; eyes composed of two groups of up to eight ocelli; they can be various colours and are sometimes velvet textured.

Proturans (Class Protura)
Very small (less than 2 mm in body length); elongate cylindrical body with no colouration; wingless; head with no eyes and no antennae; front legs are often carried forward to resemble antennae; no 'tails'.

Diplurans (Class Diplura)
Small to large (4–50 mm in body length); elongate and cylindrical body, usually with no colouration; head with thin multi-segmented moniliform antennae and no eyes; abdomen either with two 'tails' or a pair of darker coloured pincers—the similarity of these forms to earwigs is easily distinguished by the lack of eyes in diplurans.

LEFT: Pyrgomorph grasshopper, order Orthoptera, **Pyrgomorphidae**

Life Cycle, Biology and Classification

Springtails (Class Collembola)

Springtails go through a quick generation cycle that involves eggs laid in the soil, followed by five or six immature instars before the adult stage. This takes from three to five weeks allowing at least three to four generations a year. The adults continue to grow and alternate between an instar that usually does not feed but can reproduce, and one that cannot reproduce but does most of the feeding. Springtails feed mainly on microorganisms ranging from bacteria to protozoans, although some species are herbivores and scavengers. There are virtually no habitats without springtails and their density can be extremely high: from 100 000/m^2 on sub-antarctic islands, to about 2000/m^2 in the arid interior.

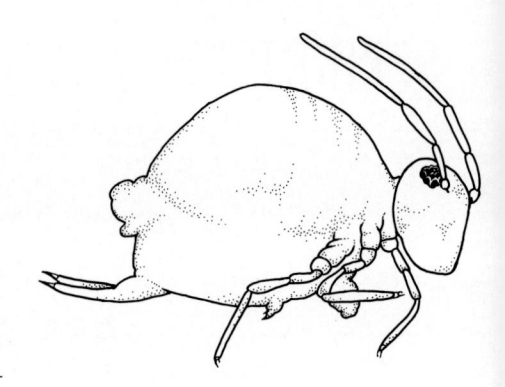

Figure 14. *Katianna* sp. (family Sminthuridae). One of the Symphypleona group of globular springtails. They are 4 mm in length and live in litter and grass. [R. Redman]

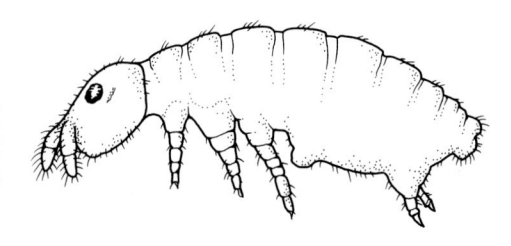

Figure 13. *Brachystomella* sp. (family Brachystomellidae). One of the Arthropleona, elongate, group of springtails which often appear in huge numbers in pasture pools after rain. It is blue, velvety and up to 6 mm in length. [R. Redman]

The 14 families are placed in three orders. The majority (1160 species in 11 families) are in the order Arthropleona which are the elongate body types (see Figure 13). Around 450 species in two families are placed in the order Symphypleona and have globular bodies and long antennae (see Figure 14). Another 20 globular species with short antennae and no eyes are in the small order, Neelipleona.

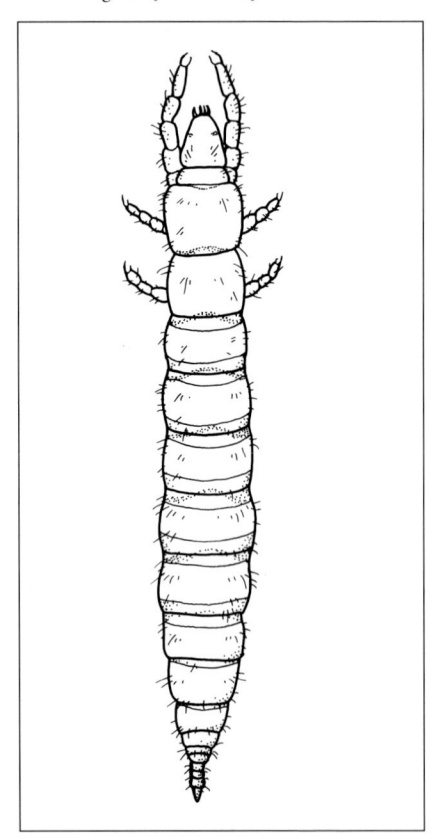

Figure 15. A typical proturan in the major family Acerentomidae, only 1.5 mm long. [after Silvestri]

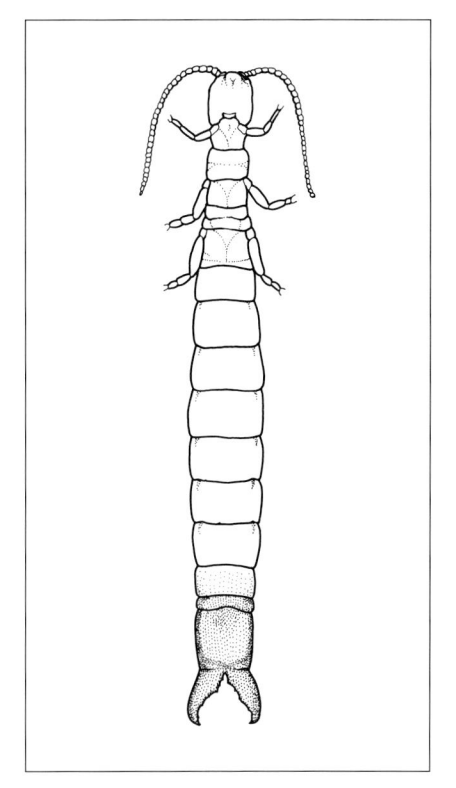

Figure 16. A dipluran, *Heterojapyx* sp. (family Heterojapygidae). Ventral view, 35 mm long. [after Tillyard]

Proturans (Class Protura)

Proturans live in the soil and leaf litter, and being less than 2 mm in length are seldom encountered without special effort (see Figure 15). Little is known about their biology. The life cycle involves an egg, several immature instars and a sexually mature adult stage. They apparently feed on fungi. The three families are difficult to distinguish and have a wide distribution, mainly in forest habitats.

Diplurans (Class Diplura)

Diplurans dwell and breed in the soil. Between egg and adult is an undetermined number of immature instars which differ from adults only by having fewer antennal segments. The adults continue moulting up to 30 times in a lifespan of up to one year. Species in some families are herbivores, but the three families with pincer-like cerci are predators, using the pincers to catch prey in and above the soil.

Twelve species in the families Campodeidae and Projapygidae have filamentous cerci. The remaining 19 species in the families Japygidae, Heterojapygidae and Parajapygidae have pincer-like cerci. The genus *Heterojapyx* (Figure 16) has large (10 mm) and pincered species in eastern Australia.

Further Reading

Condé, B. and Pagés, J. (1991), 'Diplura' (ch. 13) in *The Insects of Australia*, CSIRO, Melbourne.

Greenslade, Penelope J. (1991), 'Collembola' (ch. 11) in *The Insects of Australia*, CSIRO, Melbourne.

Imadaté, G. (1991), 'Protura' (ch. 12) in *The Insects of Australia*, CSIRO, Melbourne.

BRISTLETAILS ORDER ARCHAEOGNATHA
SILVERFISH ORDER THYSANURA

Bristletails (Order Archaeognatha)
World families: 2; species: 350 **Australian families:** 1; species: 7

Silverfish (Order Thysanura)
World families: 4; species: 370 **Australian families:** 2; species: 28

Characteristics of Orders
The principal and simplest difference between these similar orders is the large compound eyes of bristletails. Their straight 'tails' also differentiate bristletails from silverfish which have more outward pointing 'tails'.

Bristletails (Order Archaeognatha)
Elongate, rounded, wingless body, with fine scales; very large compound eyes, meeting in the middle; long, body-length filiform antennae; long seven-segmented maxillary palps; three 'tails' or cerci at the tip of the abdomen, the middle cercus is twice the length of the outer two and all three point straight back. Bristletails can jump by flexing the body.

Silverfish (Order Thysanura)
Elongate, flattened, wingless body, with or without fine scales; compound eyes are small or absent; filiform antennae which are shorter than the body; short five-segmented maxillary palps; three 'tails' or cerci, with the outer two pointing away from the body. Silverfish cannot jump.

Life Cycle

Bristletails (Order Archaeognatha)
Reproduction does not involve mating as such, as the male simply deposits sperm for the female to pick up. Eggs are laid in batches of 2 to 30 in crevices. Often the eggs enter diapause and do not hatch until the next good season. The larvae resemble the adults and go through about eight or nine instars before becoming sexually mature.

Silverfish (Order Thysanura)
Similar indirect mating as in bristletails, but the eggs usually hatch within two months. The larvae are also shaped like the adult, though they undergo 11 to 14 instars before reaching sexual maturity. Silverfish continue moulting as adults and live up to four years, laying fresh batches of eggs after every moult.

Biology and Classification

Bristletails (Order Archaeognatha)
All seven Australian species are in the family Meinertellidae. Both nocturnal and diurnal species live in damp habitats such

BRISTLETAILS AND SILVERFISH

as under bark or in leaf litter. The largest species (18 mm in length), *Allomachilis froggatti*, sometimes aggregates in large numbers on sea-sprayed coastal cliffs from Qld to southwest WA (see Figure 17).

What little is known of the diet of this order includes detritus, lichens and algae. When disturbed, bristletails can snap their bodies to propel themselves up to 10 cm into the air.

Silverfish (Order Thysanura)

The two Australian families are easily separated: the Nicoletiidae, with 12 species, are without eyes; while the 16 species of Lepismatidae have small compound eyes.

Nicoletiids are generally very small (under 7 mm in length) and many live in association with ants and termites which makes them unlikely to be encountered. Most are herbivores, though an omnivorous diet is possible.

Lepismatids are the more typical silverfish. Among them are six species found almost worldwide and introduced into Australia. They are the silvery-white insects that live in dwellings and feed on substances such as flour and old glue. Their fine scales make them very slippery, as the scales shed in their captors' grip. *Ctenolepisma* is the main genus in this group. Native lepismatids live in litter and among termites and ants. Their distribution is very wide, including a desert species which can absorb enough moisture directly from the air to survive.

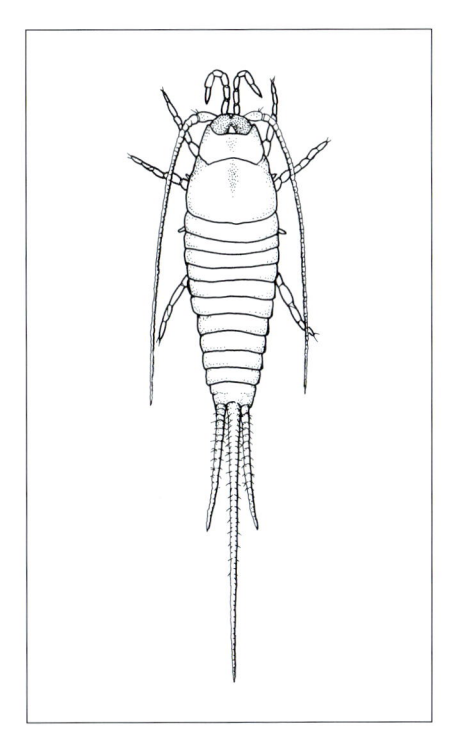

Figure 17. A typical bristletail, *Allomachilis* sp., 14 mm. [R. Redman]

Silverfish, *Ctenolepisma* sp., **Lepismatidae** [10 mm long]

Further Reading

Smith, G.B. and Watson, J.A.L. (1991), 'Thysanura' (ch. 15) in *The Insects of Australia*, CSIRO, Melbourne.

Watson, J.A.L. and Smith, G.B. (1991), 'Archaeognatha' (ch. 14) in *The Insects of Australia*, CSIRO, Melbourne.

Womersley, H. (1939), *Primitive Insects of South Australia*, Government Printer, Adelaide.

MAYFLIES
ORDER EPHEMEROPTERA

World families: 23; species: 2500 Australian families: 9; species: 84

Characteristics of Order

Incomplete life cycle; medium-sized (average wingspan is about 15 mm); most with two pairs of membranous wings with the hind wings very much smaller than the forewings and absent in a few. Head with mandibulate mouthparts in the aquatic nymphs, but reduced and non-functional in adults; very large eyes and three ocelli; very short (about head length) filiform antennae; abdomen with usually three, rarely two, long filaments (cerci), rarely less than twice the body length. Aquatic nymphs are similar shape to adults minus wings and with abdominal gills which look like fine leaves along both sides of the abdomen.

Features of mayflies that distinguish them from other aquatic orders such as stoneflies (Plecoptera), caddisflies (Trichoptera) and alderflies (Megaloptera) include their very short antennae, vertically held wings, very much smaller hind wings and, in most mayflies, three very long tail filaments.

Life Cycle

Mayflies are fascinating for being the only insects with an extra pre-adult, winged 'sub-imago' stage, and for their extremely short adult life spans. Most adults live only a few hours to a day, during which time the males swarm at a time of day and habitat characteristic of that species. Mating occurs in flight and eggs are mainly laid on the water surface and sink. Several hundred to several thousand eggs are laid per female which hatch two to three weeks later. The nymphs live on the bottom of streams and lakes where they undergo up to 50 moults, with 20 being more usual (see Figure 18). Emergence to sub-adult and adult is synchronised in many species to allow the short lived adults maximum time to mate and lay eggs. In these normally colder country species, mating swarms are quite spectacular and

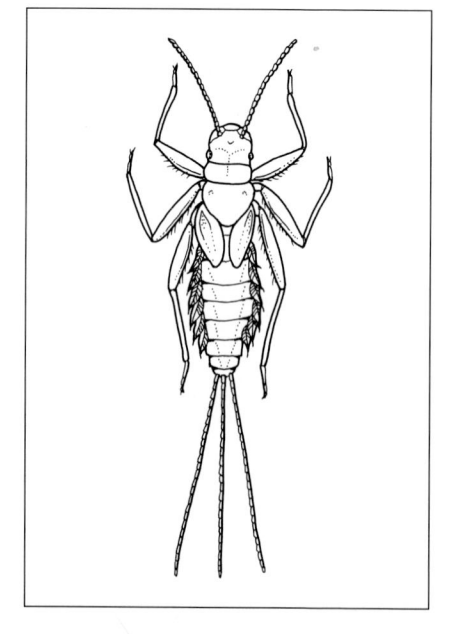

Figure 18. Typical aquatic mayfly nymph. Genus *Atalophlebia* (family Leptophlebiidae). [adapted from R.J.Tillyard]

Atalophlebia sp., **Leptophlebiidae** (10 mm long)

occur once a year. In the tropical species there can be several generations a year and these are rarely synchronised.

Biology

Each species prefers a different type of water body, depending on water temperature, flow rate, light, and chemical content. Most species prefer cold, clear, flowing streams and therefore they are most diverse in the mountain streams of eastern Australia and Tasmania. Only a few species live in the interior rivers and in SA and WA.

The nymphs feed by scraping algae and detritus off stones and leaves underwater. In the final few instars, they lose their functional mouthparts and the alimentary canal empties and enlarges to become a buoyancy device. Filled with air, this helps in egg laying on the surface, as well as in flight.

Classification

The family Leptophlebiidae contains 54 out of the 84 species. As nymphs they can generally be distinguished by their leaflike, paired gills. *Atalophlebia* is the largest genus with 18 species found in many habitats. The genus *Jappa* has gravel-burrowing nymphs and is also widespread.

Other families usually have a single pair of gills per segment. The adult features used to identify many families

Jappa sp., **Leptophlebiidae** (15 mm long)

Cloeon sp., **Baetidae** (14 mm long)

are very obscure, using details of wing venation and internal genitalia.

Adult males of the 13 species of Australia's second largest family, the Baetidae, have the eyes divided completely into a large lower part and smaller upper part, and some genera such as *Cloeon* lack the hind wings.

In all six species of Caenidae the hind wings are also absent. The remaining five families have only nine species between them.

Further Reading

Campbell, I.C. (1986), 'Life histories of some Australian siphlonurid and oligoneuriid mayflies (Insecta: Ephemeroptera)', *Australian Journal of Marine Freshwater Research*, vol. 37, pp. 261–88.

Peters, W.L. and Campbell, I.C. (1991), 'Ephemeroptera' (ch. 16) in *The Insects of Australia*, CSIRO, Melbourne.

DRAGONFLIES AND DAMSELFLIES
ORDER ODONATA

World families: 26; species: 5000 **Australian families:** 17; species: 302

Characteristics of Order

Incomplete life cyle; medium to large (20–150 mm in body length); thin, elongate, cylindrical insects with large heads and two pairs of membranous wings; hind wings broader than forewings, at rest spread horizontally in dragonflies and usually vertically in damselflies; large to very large compound eyes (meeting in the middle in most dragonflies); three visible ocelli; antennae are minute and less than five-segmented; biting mouthparts with toothed mandibles; legs are short with clawed, three-segmented tarsi adapted for grabbing prey.

Larvae are aquatic; only slightly similar to wingless adults, with the body broader, flattened and shorter; head is equipped with a 'mask', a modified labium which can spring forward to catch prey; gills are present at the tip of the abdomen only, concealed in a squat pyramid shape in dragonflies and three exposed fine leaves in damselflies.

Differences from other aquatic orders such as mayflies, alderflies and stoneflies are treated in more detail in those chapters. The very typical form of dragonflies and damselflies with their tiny, pointed antennae are good distinguishing characteristics for this order.

Life Cycle

Mating can be while stationary or during flight, involving unique contortions The male clasps the female on the neck with the tip of his abdomen, while the female's abdomen clasps the male's second and third abdominal segments (unlike most insects, male Odonata have a set of modified genitalia on these segments). This mating configuration is known as the 'wheel' position. They can fly together like this or in the 'tandem' position, with only the neck connection. The female then deposits eggs in the water while guarded by the male.

The larvae are predatory and live in all types of freshwater (see Figures 19–21). After 10 to 15 moults, the final larval instar crawls out onto vegetation or a rock where the adult emerges. Some species which utilise temporary water bodies can become adults in two months, but a year or even two is more usual. Species with terrestrial larvae have been recorded in Qld.

Biology

Dragonflies are territorial, diurnal and showy insects, displaying complex behaviour patterns. Males tend to live near water, guarding a hunting and mating territory. They often have a favourite perch from which to survey the area with their huge eyes which can have up to 30 000 facets. They hunt small insects in flight and sometimes settle to consume them.

The females tend to roam further away from water and are often different from the males, especially in colour which tends to be duller. Both sexes are very agile fliers. Their wings do not couple in flight and this independent motion helps them to manoeuvre very tight turns and even to suddenly fly backwards. Short

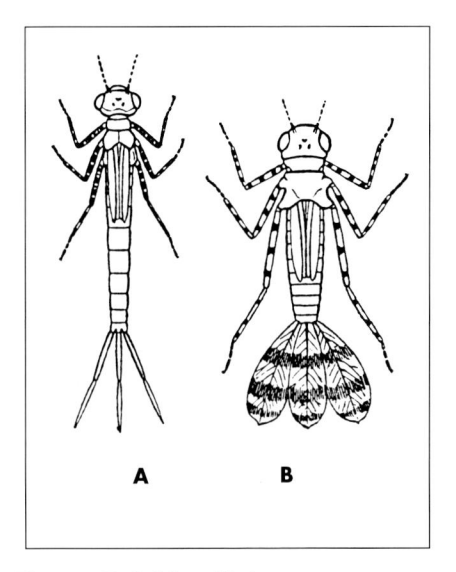

Figure 19. Typical damselfly (Zygoptera) larvae: (a) family Coenagrionidae; (b) family Megapodagrionidae.

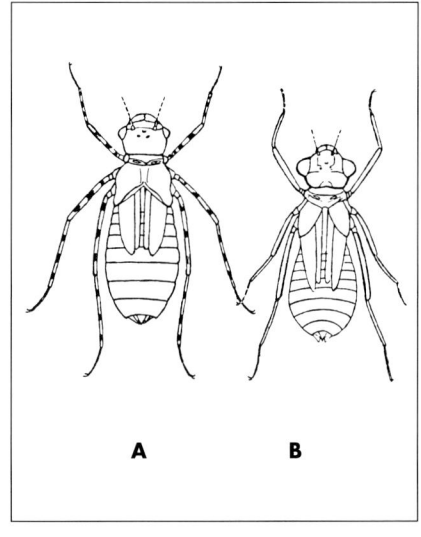

Figure 20. Typical dragonfly (Anisoptera) larvae: (a) family Corduliidae; (b) family Libellulidae. [R.J. Tillyard]

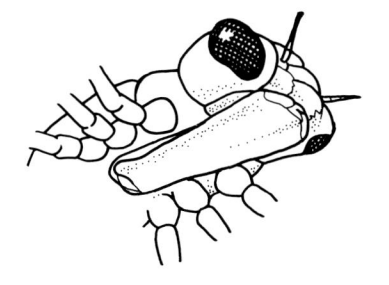

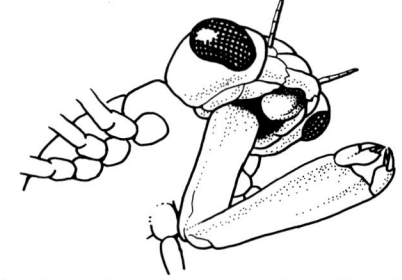

Figure 21. The modified labium mouthparts (mask) used to catch prey by the odonate larvae. [R. Redman]

bursts of speed show amazing acceleration and can attain 40 kph.

As both adults and larvae eat large numbers of mosquitoes, they are regarded as beneficial insects.

Classification

Two suborders divide the stout dragonflies from the delicate damselflies. The differences are simple:

Damselflies (suborder Zygoptera):
- Both pairs of wings are slender, tapering near the body, almost equal in length and not held horizontal at rest.
- Larvae have three leaf-like gills at the tip of the abdomen.

Dragonflies (suborder Anisoptera):
- Hind wings are broader than forewings, and wings are always held horizontal at rest.
- Larvae do not have leaf-like gills.

Damselflies (Suborder Zygoptera)

There are 11 Australian families with 107 species. Classification of adults uses differences in wing venation, and gill and mask details in the larvae. Most of the families have few species. The largest family is the Coenagrionidae with 30 species, including the delicate *Ceriagrion aeruginosum*. The Amphipterygidae has some of the biggest damsels, with eight species in the genus *Diphlebia* found mainly in the east of Australia and near fast, clear streams. The family Lestoideidae has only two species including the yellow *Lestoidea conjuncta* in north Qld.

Lestoidea conjuncta, **Lestoideidae** (28 mm long)

Ceriagrion aeruginosum, **Coenagrionidae** (34 mm long)

Diphlebia euphaeoides, **Amphipterygidae** (48 mm long)

Female *Diplacodes haematodes*, **Libellulidae**
(38 mm long)

Male *Diplacodes haematodes*, **Libellulidae**
(40 mm long)

Male *Orthetrum caledonicum*, **Libellulidae**
(45 mm long)

Austrogomphus prasinus, **Gomphidae**
(40 mm long)

Dragonflies (Suborder Anisoptera)

There are six Australian families with 195 species. Classification is again mainly based on differences in wing venation. Two families stand apart by having their eyes noticeably separated rather than meeting in the middle: Petaluridae, with four species, includes *Petalura ingentissima*, one of the world's largest dragonflies with a 160-mm wingspan; and Gomphidae, with 38 mainly black-and-yellow species like *Austrogomphus prasinus* found in Qld.

The largest family with 55 species is the Libellulidae, with many red-coloured species and a tendency for shorter, stouter bodies than other families. The all-red males of *Diplacodes haematodes* are typical of this family and a good example of the sexual dimorphism common in dragonflies, as the female looks nothing like the male. While blue is a commoner colour among the damsels than the dragonflies, the blue males of *Orthetrum caledonicum* are a common sight in most parts of the mainland.

Further Reading

Corbett, P.S. (1983), *A Biology of Dragonflies*, Classey, Faringdon, UK.

Watson, J.A.L. and O'Farrell, A.F (1991), 'Odonata' (ch. 17) in *The Insects of Australia*, CSIRO, Melbourne, pp. 294–310.

Watson, J.A.L., Theischinger, G. and Abbey, H. (1991), *The Australian Dragonflies. A Guide to the Identification, Distribution and Habitats of Australian Odonata*, CSIRO, Melbourne.

STONEFLIES
ORDER PLECOPTERA

World families: 15; species: 2000 Australian families: 4; species: 196

Characteristics of Order

Incomplete life cycle; small to large (wingspan: 10–110 mm); body elongate, flattened and soft; forewings and hind wings are membranous, held flat or almost curved around the body at rest and with the hind wings broader than the slightly longer forewings; mandibulate mouthparts; legs are long, forward pointing at rest, with three-segmented tarsi; abdomen with two cerci of various lengths.

Larvae are aquatic, similar to the adults with short 'flaps' in lieu of wings, and with gill tufts along the sides of the body, the tip of the abdomen, and the edges of the legs.

Life Cycle

Adults live and mate near water, and the females either lay eggs on the water surface or submerge to lay on stones. The 100 to 1000 eggs take up to a year to hatch, and the larvae are usually slow growing (see Figure 22). There are 10 to 15 moults to the adult stage, which together take from a few months in the tropics to a few years in the coldest streams. The larvae have a range of diets from algae, to leaves, detritus and even predation.

Biology

Unlike the extremely short-lived mayflies, stonefly adults do feed and live from a few days to a few months. They are always found near the fresh

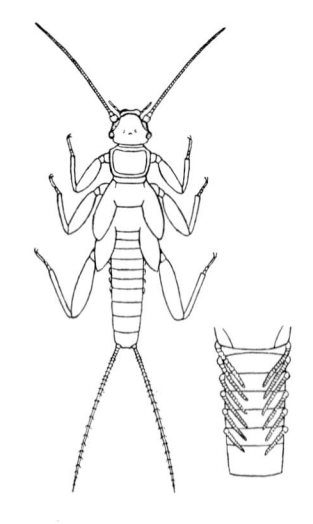

Figure 22. Typical aquatic larva (family Eustheniidae) with a close-up view of the gills on the underside of the abdomen. [from R.J. Tillyard

STONEFLIES

Dinotoperla sp., **Gripopterygidae** (16 mm long)

water they breed in, feeding on algae, rotting wood and other detritus. Smaller species will crawl rather than fly when disturbed, but most species can fly, and the larger ones often live high up in trees, occasionally coming to light. The majority of stonefly species prefer cool, clear streams at high altitudes, and this is reflected in their distribution. Tasmania has 75 species, the rest are mainly in eastern Vic, NSW and eastern Qld. Only a few species are found in SA and the south-west of WA.

Classification

The four families found in Australia are restricted to the southern hemisphere. All Australian species are endemic, and over 70% are in the family Gripopterygidae. Adult identification characteristics are principally wing venation and the subtle marks of the remnant larval gills. The larvae of stoneflies are the stage most likely encountered by people searching for water insects, and their characteristics, mainly the gill arrangements, are a little easier to discern. The Gripopterygidae stand out by being the only plecopterans with a tuft of gills between the cerci. While some stoneflies can be quite colourful, with green, red and even purple wings, most adult Gripopterygidae are of mottled red or brown appearance. The 35 species of *Dinotoperla* are typical of the adults of this family. The next largest family, the Notonemouridae with 29 species, can be recognised as adults by being the only stoneflies not to have cross (vertical) veins in the outer half of the wings.

Further Reading

Theischinger, G. (1991), 'Plecoptera' (ch. 18) in *The Insects of Australia*, CSIRO.
Theischinger, G. and Cardale, J.C. (1987), 'An illustrated guide to the adults of the Australian stoneflies (Plecoptera)', *Australian Division of Entomology Technical Papers*, (CSIRO) vol. 26, pp. 1–83.

COCKROACHES
ORDER BLATTODEA

World families: 6; species: 4000 Australian families: 5; species: 428

Characteristics of Order

Incomplete life cycle; small to large (3–70 mm in length); broad, flattened body with the pronotum forming a shield usually overhanging the body on the sides and partly the head; wings, when present, are membranous with toughened forewings which overlap left over right; head has mandibulate mouthparts, pointing downwards and with the ocelli reduced to two lenseless points between the eyes; legs are long, spined and adapted for running.

The general form of cockroaches varies little. Therefore, the image of the introduced city dwelling species helps to place the majority of the native bush species in this order.

Life Cycle

Sexual dimorphism is common. Usually this consists of the females being wingless in winged species. In a few species, the differences are enough to confuse the sexes for different species.

In the majority of cockroaches, females produce a hard egg case known as an ootheca. This is carried around for some time before being deposited either in ground litter or onto a tree trunk. Between 12 and 40 eggs are inside each case which splits to release the young. Normally the nymphs fend for themselves, but some species in the family Blaberidae rear their young in underground chambers. The nymphs look like colourless, wingless adults and undergo about six moults before developing wing buds. In the winged species (about 40% of species), wings appear on the last moult. Cockroaches develop slowly, taking several years in some cold climate species, with a year being normal for the introduced city species.

Biology

Australian cockroaches are widely distributed, being adapted to both wet and arid regions. The highest diversity is found in the warmer north. While some native species forage on bushes in sunlight, most species are nocturnal, hiding during daylight in crevices, under logs and in burrows. The food of native species is largely unknown but it is likely to be mainly the detritus associated with surface leaf litter. Some species possess gut protozoa to break down cellulose and feed on the rotting logs they inhabit. The introduced species will eat almost any organic matter in the houses they infest. At least 10 species are found in caves, with at least two species being blind and restricted to caves at Undarra Lava Tubes and on the Nullarbor Plain.

Nymphs and adults are prey to many birds, mammals and other arthropods; though species in the family Blattidae often produce very foul smelling and often poisonous secretions in defence.

Classification

Five of the six families of cockroaches are found in Australia. The Blattidae contains half the species, including most of the 10 or so introduced species. The majority of Australian genera in this and other families are endemic.

Cosmozosteria sp., **Blattidae** (26 mm long)

Family characteristics revolve largely around differences in the shapes of the male and female genitalia and external subgenital plates and are therefore difficult to separate in the field.

Family Blattidae

While most of the 200-plus species are bush cockroaches, the best known members of this family are the introduced pests. The large (40 mm) shiny red American cockroach, *Periplaneta americana* and the so-called Australian cockroach, *P. australasiae*, which originated in Asia, are an integral part of domestic life. *P. australasiae* can be recognised by a yellow band across the base of the pronotum and yellow slashes along the base of the forewings. Their effective diurnal hiding and the ability to fly into new territory make them all but impossible to eradicate. Among the bush cockroaches are many large wingless genera. These fat sometimes colourful cockroaches are often diurnal and clamber about on the bushes they feed on. Many can emit a powerful defensive chemical smell. *Polyzosteria* and *Cosmozosteria* have some very striking, flightless, sometimes metallic coloured species.

Australian cockroach, *Periplaneta australasiae*, **Blattidae** (30 mm long)

Family Blattellidae

The 140 or so species are typical bush cockroaches. In many genera, the males are winged; the females having only a reduced wing cover plate looking like an extra segment of the pronotum. Most are nocturnal and some come to lights at night, looking like small delicate versions of the domestic roaches. Some of the 42 species in the largest genus *Balta* scamper about bushes during the day, as does the very attractive and widespread *Ellipsidion australe*.

This family also has another well-known pest species, the German

COCKROACHES

Ellipsidion australe, **Blattellidae** (15 mm long)

Molytria sp., **Blaberidae** (30 mm long)

cockroach, *Blattella germanica*. In Germany it is called the 'Russian cockroach'; in Russia, 'the German'; and is most likely to have originated in Asia.

Family Blaberidae

Some biologically fascinating species are found in this family of around 70 species. Several genera live in rotting logs. The genus *Panesthia* has large, heavily armoured species that live in family groups which apparently nurture their young. The adults have very hairy rather than spiny legs.

The most bulky cockroach in the world, *Macropanesthia rhinoceros* (Figure 23), is wingless and lives in tunnels in the soil in northern Qld. The females feed the young on dead leaves scavenged on the surface. It can reach 70 mm in length, weigh up to 20 g, and can emit a hissing noise when threatened.

Another group of genera in this family has very flat species which live under bark. Species of the genus *Molytria* are among the largest. In the genus *Laxta*,

Figure 23. *Macropanesthia rhinoceros* (family Blaberidae) 70 mm long. [G. Thompson]

males are winged and flat, while females are wingless and even thinner, resembling a very thin woodlouse. They have been called 'trilobite roaches' in the sandstone woodlands of Sydney.

Other Families

There are two other, very small, families. The two species of the Nocticolidae are from north Qld where one lives inside the Chillagoe Caves. The Polyphagidae, with only five species, are found in the arid interior.

Further Reading

Guthrie, D.M. and Tindall, A.R. (1968), *The Biology of the Cockroach*, Edward Arnold, London.
Rentz, D.C.F. (1995), *Australian Orthopteroid Insects*, University of NSW Press, Sydney.
Roth, L.M. (1991), 'Blattodea' (ch. 19) in *The Insects of Australia*, CSIRO, Melbourne.
Rentz, D.C.F. (1996), *Grasshopper Country*, University of NSW Press, Sydney.

TERMITES
ORDER ISOPTERA

World families: 7; species: 2300 **Australian families:** 5; species: 348

Characteristics of Order

Incomplete life cycle; small to medium (3–18 mm in body length); elongate, pale bodies with several body forms, or castes, which live in social colonies; short cerci; wings are present only in reproductive castes and are longer than the body, held flat over the body with the anterior veins much thicker than the others; biting mandibulate mouthparts which are enlarged in soldier castes of many species (some soldier castes have a large snout pointing forward from the head); antennae are moniliform and about as long as the head.

As both termites and ants live in multicaste colonies, and termites are confusingly referred to as 'white ants', it is important to differentiate between the two. Ants are a family of the wasp order, Hymenoptera, and as such have a very pronounced 'waist', a more armoured and pigmented body, antennae with distinct elbows and the workers have compound eyes. Termite workers have no waist, no eyes, straight antennae and soft, pale bodies.

Life Cycle

Mature colonies of termites produce a reproductive caste once or twice a year. These are the only winged and non-sterile termites which, on very humid nights usually, fly away from the nest to start new colonies. Wings are shed after a short nuptial flight and the individual pairs then mate and construct a shelter in which the first eggs are laid. When this first brood of sterile workers matures and is able to feed the female and enlarge the nest,

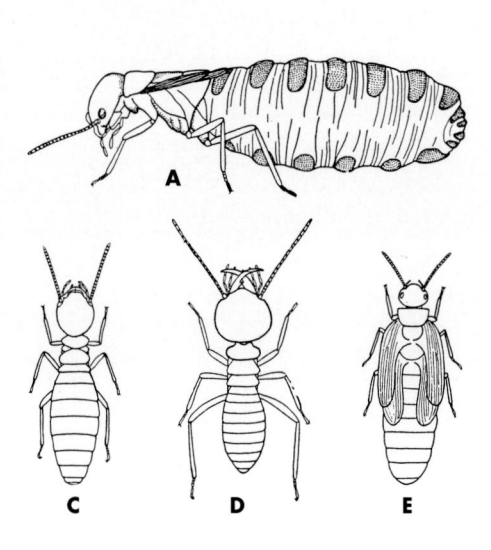

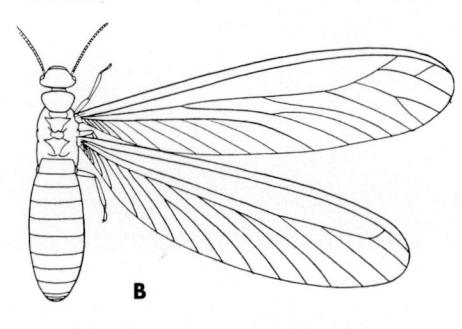

Figure 24. The castes of termites illustrated by *Hermitermes silvestri* (family Termitidae): (a) queen, 12 mm; (b) winged male, 10 mm; (c) worker, 4 mm; (d) soldier, 5 mm; (e) nymph, 7 mm. [after A. Tonnoir]

the queen settles into a life of almost continuous egg laying. In some species, a queen can lay up to 2000 eggs per day. These colony-founding reproductives can live many years, even decades. There are records of individual colonies remaining active for more than 50 years, and 100 years or more is likely. After several years of growth, a new colony is ready to produce its first reproductives, again sending these out on annual dispersal flights.

Biology

Termites live in colonies founded by a single reproductive pair. Large earthen mound nests are made by only a small number of species, with most termite species living either inside the timber on which they feed or in underground nests. The colonies contain from several hundred to 2 million individuals. The great majority of these are the worker caste which builds and repairs the nest, forages for food and feeds the queen. Each nest also has a soldier caste of larger, more heavily sclerotised termites with either large mandibles or a snout which shoots out a sticky secretion to ensnare an attacker Soldiers represent only a few percent of the colony and are often seen at a damaged part of the nest, protecting the workers while repairs are made (see Figure 24). The queen and original reproductive male are usually hidden deep inside the nest and the abdomen of this female is often distended with eggs to many times its normal size.

Termites feed mainly on cellulose which they obtain by eating wood, bark, grasses, leaves, humus, fungi and even herbivore droppings. In turn, termites are an important food item for mammals such as the echidna and numbat, for many lizards and snakes, and many invertebrates. Termites are a fattening food source for many birds and bats during their brief mating flights. While about 20 of the 348 Australian species are considered to be pests of human constructions, most species have an important role to play in the recycling of nutrients and in the formation of soils, especially in the tropics.

Classification

Of the five families represented in Australia, only the family Mastotermitidae is endemic. It has only one large primitive species found in the tropics, *Mastotermes darwiniensis*, which is a serious pest of timber, crops and even plastics. It normally lives in nests within trees and logs and is the only termite with five-segmented tarsi (four being normal for all others).

The other four families can be roughly separated using the differences described below (comparing the soldier caste only):

- If the cerci at the tip of the abdomen are four or five segmented, then the family is Termopsidae. Five species with mandibulate soldiers are included, all of which live in soft rotting wood from Tas to Qld.
- If the cerci are stout and only two segmented, then look for the fontanelle on the front of the head: it is like a 'third eye', round and raised. The fontanelle places the termite in one of two families: Rhinotermitidae or Termitidae. Its absence places it in the family Kalotermitidae which has 46 species with mandibulate soldiers, most of which live inside the wood they attack. Some attack wooden power poles and some species in the largest genus *Cryptotermes* are pests of home timbers. The genus *Neotermes* has the largest soldiers of all the termites. These live in galleries in living timber and reach 12 mm in length.
- If a fontanelle is present, then look at the shape of the pronotum (first part of the thorax from above). It can be either a simple rounded, flat shape, or saddle shaped with raised lobes on the outer edges. A simple flat pronotum belongs to the family Rhinotermitidae which has 30 species with mandibulate soldiers and usually underground nests, with some species building mounds. The genus *Coptotermes* has many economically important pest species and is widely distributed. Some tropical species build domed and conical mounds up to 3 m high.
- If a fontanelle is present, or elongated into a snout, and the pronotum is saddle shaped, then the family is the Termitidae, with 266 species in two distinct subfamilies:

    The subfamily Termitinae have mandibulate soldiers, including several genera with long, thin mandibles which are used for snap-ping rather than chewing. A good example is the 3-mm long *Hapsidotermes.* The majority of species in this subfamily are grass and detritus feeders and are most abundant in warmer areas including central Australia. Most live underground, but some make nests high up in trees. Others invade and share the mounds of other termites and a few are well-known mound builders. Among these are the so-called 'magnetic' termites of Cape

Giant northern termite workers and a soldier (right), *Mastotermes darwiniensis* **Mastotermitidae** (12 mm long)

LEFT: Mounds of magnetic termite, *Amitermes* sp., **Termitidae** (average 2.5 m tall)

BELOW: Soldier of *Hapsidotermes maideni*, **Termitidae** (4 mm long)

ABOVE: Mound of spinifex termite, *Nasutitermes triodiae*, **Termitidae** (5 m tall)

RIGHT: Soldiers of spinifex termite, *Nasutitermes triodiae*, **Termitidae** (6 mm long)

58

York and the Top End, *Amitermes laurensis* and *A. meridionalis*, which make tall flat mounds always aligned north–south.

Members of the subfamily Nasutitermitinae are easy to recognise because all soldiers have snouted heads. Colonies of the grass-feeding *Nasutitermes* can contain millions of individuals. *N. triodiae* from northern Qld and NT build the largest mounds which can be more than 7 m tall. The genus *Tumulitermes* has over 50 species, and mound builders like *T. hastilis* are very abundant in the warmer inland.

Further Reading

Hadlington, P. (1987), *Australian Termites and other Common Timber Pests*, University of NSW Press, Sydney.

Watson, J.A.L. and Gay, F.J. (1991), 'Isoptera' (ch. 20) in *The Insects of Australia*, CSIRO, Melbourne.

PRAYING MANTIDS
ORDER MANTODEA

World families: 8; **species:** 1800 **Australian families:** 3; **species:** 162

Characteristics of Order

Incomplete life cycle; medium to large (10–120 mm in length); predators with raptorial forelegs which have one or two rows of spines along the tibia; mandibulate mouthparts; large eyes set in a very mobile head; short to moderate length, filiform antennae; forewings are protectively hardened (tegmina); hind wings are membranous, sometimes reduced or absent.

Life Cycle

The female mantid lays up to several hundred eggs in a distinctive case called an ootheca. Accessory glands produce this as a foam which quickly hardens. It is attached to branches and trunks of plants, or placed on the ground under logs and stones. The hatchlings develop through a series of nymph stages which resemble wingless adults. In colder climates, one generation takes about a year to develop; but in the tropics, two generations can overlap in this time.

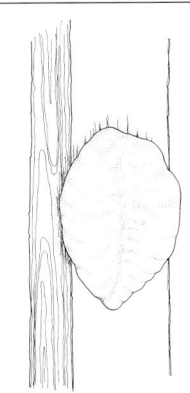

Figure 25. A typical egg case or ootheca of mantids in the genus *Archimantis*, 50 mm across [R. Redman]

Tenodera sp., **Mantidae** (70 mm long)

Biology

Mantids are general rather than specific predators, eating mainly insects of a size they can overpower. Most mantids hunt sitting on vegetation, but a few genera have groundrunning species. Some species hunt on particular sites such as flowers, tree trunks or rocks, and in these cases are often cryptically coloured to blend with their background. While most species are green or brown, there are some species that change colour over time, especially those that dwell in grasses that seasonally dry from green to brown.

Natural enemies of mantids include parasitic wasps which parasitise their eggs, and various arthropods like ants and crickets which attack the young nymphs. Older mantids tend to be protected by their moderate size and cryptic habits, although some fall prey to birds. A few species also resort to warning colour flashes, unpleasant chemicals or aggressive displays to ward off predators.

Classification

Only three of the eight world families are represented in Australia: the Hymenopodidae which has one species found in north Qld; the Mantidae which is the largest family with 116 species; and the Amorphoscelidae which includes the remaining species.

Family Mantidae

This family can be distinguished by the presence of an extra row of spines on the lower external margin of the fore tibia, compared to a single more central row in the other families. The genus *Tenodera is* a common, widely distributed inhabitant of bush and suburb greenery. It has species up to 10 cm long, which are usually brown and winged. Even larger similar species are found in the genus *Archimantis* and these differ by having females with wings extending only halfway along the abdomen (brachypterous). In the warmer parts of Australia, the genus

PRAYING MANTIDS

Orthodera ministralis, **Mantidae** (40 mm long)

Neomantis australis, **Mantidae** (18 mm long)

Nesoxypilus albomaculatus, **Amorphoscelidae** (14 mm long)

Paraoxypilus sp., **Amorphoscelidae** (20 mm long)

Gyromantis sp., **Amorphoscelidae** (25 mm long)

Orthodera is common with several green species characterised by a blue spot on the inside of the front legs. The tropics are home to the distinctive delicate genus, *Neomantis*, which hunts among bushes, even at night.

Family Amorphoscelidae

This family includes 45 species which live on tree trunks and on or near the ground. Many have very effective cryptic colouration; and in the majority of species, the females are either wingless or brachypterous, sometimes resembling very robust ants.

61

Nesoxypilus albomaculatus from the tropics, is a very good example. Species of *Gyromantis*, characterised by spines behind the eyes, live on tree trunks and can resemble, for example, the paper bark of *Melaleuca*. The ground- and bark-running *Paraoxypilus* are sometimes found in a black (melanic) colour phase in areas of bushfire-blackened trees.

Further Reading

Balderson, J. (1991), 'Mantodea' (ch. 21) in *The Insects of Australia*, CSIRO, Melbourne.

Rentz, D.C.F. (1995), *Australian Orthopteroid Insects*, University of NSW Press, Sydney.

Rentz, D.C.F. (1996), *Grasshopper Country*, University of NSW Press, Sydney.

EARWIGS
ORDER DERMAPTERA

World families: 10; **species:** 1800 **Australian families:** 7; **species:** 63

Characteristics of Order

Incomplete life cycle; small to large (5–50 mm in length); forceps-like structures at the tip of the abdomen are present in most species; many are wingless but, when present, the forewings are protectively hardened and short, covering only the thorax; the hind wings are membranous and fanlike with complex folds to fit under the small covers; mandibulate mouthparts; short to moderate, filiform antennae; generally elongate, flattened, smooth to shiny body; predominantly brown to black in colour.

While a small order and poorly represented in Australia, earwigs are very distinctive and common in suburban backyards and homes. The name has attracted the myth of a habit of crawling into people's ears with dire consequences. The origin of 'earwig', however, is more likely to be from the term 'ear-wing', due to the shape of the hind wings.

Life Cycle

Earwigs exhibit the rare trait in insects of maternal instinct. In many species, there is also a ritual courtship display by males before mating. The sexes can be distinguished by the shape of the forceps: the male's are curved while the female's are straight with slightly inward pointing tips. Females will lay from 20 to 80 eggs in batches deposited in burrows or natural crevices. The eggs are then guarded from predators and cleaned of fungi until hatching some two to three weeks later. The young are very similar to the adults but are paler; in winged species, the wings take four or more moults to develop. For the first two moults, the female continues guarding the young. However, after this time, they need to scatter or they face the possibility of being eaten by her.

Biology

While many species are adapted for burrowing, the majority of earwigs are found under timber, stones and other objects rather than in elaborate burrows. The forceps are mainly used defensively, curved upwards for a menacing effect, but are also used to capture and carry prey. Most species are omnivorous, with the majority of their diet being live or decaying plant matter, followed by dead insects and other organisms, and some live insect prey.

Cranopygia sp., **Pygidicranidae** (22 mm long)

Classification

Many species are either rare or live a well-hidden existence.

The family Forficulidae has three native species plus the introduced European earwig *Forficula auricularia* which is now a pest in eastern Australian gardens (see Figure 26).

Chelisoches sp., **Chelisochidae** (25 mm long)

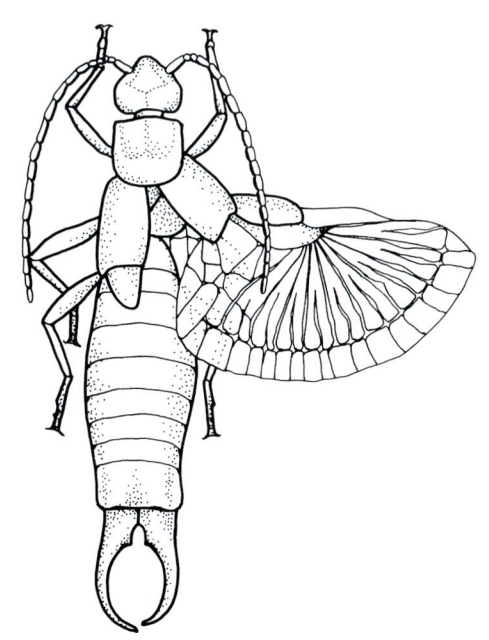

One of the world's largest earwigs is included in the wingless family Anisolabididae: *Titanolabis colossea* lives in the wet forests of the east coast and can reach 55 mm in length. The Labiduridae contains the most species throughout the continent and includes the common earwig, *Labidura truncata*. Members of the family Pygidicranidae are nocturnal, very flat, and have relatively straight forceps. *Cranopygia* is a handsome species found in Qld. The family Chelisochidae contains some colourful species in northern Australia.

Further Reading

Rentz, D.C.F. and Kevan, D.K.McE. (1991), 'Dermaptera' (ch. 23) in *The Insects of Australia*, CSIRO, Melbourne.

Figure 26. European earwig, *Forficula auricularia* (family Forficulidae), 18 mm long. [after Chopard]

CRICKETS AND GRASSHOPPERS
ORDER ORTHOPTERA

World families: 28; **species:** 20 000 **Australian families:** 14; **species:** 2827

Characteristics of Order

Incomplete life cycle; small to very large (5–100 mm in body length); enlarged hind legs which are usually modified for jumping; wings, when present, with forewings usually protectively toughened and always smaller and narrower than the broad membranous hind wings which fold like a fan; chewing mandibulate mouth-parts; generally large eyes; short to very long antennae; many have stridulatory files on either the wings or the abdomen and/or legs that produce a characteristic 'song'.

The general appearance of grasshoppers and crickets makes this order difficult to confuse with any other.

Life Cycle

In such a large group the life cycle of Orthoptera is varied. In the majority of species, the sexes find each other by the male producing calls by stridulation. Mating involves the transfer of one or more sperm capsules (spermatophores). Crickets usually lay their eggs singly, inserting them into the soil or sometimes into vegetation. Grasshoppers do not have the long tubular ovipositor of crickets and therefore tend to dig the tip of the abdomen into the soil and lay groups of 10 to 200 eggs.

Nymphs are wingless versions of the adults with, at first, disproportionately large heads. Many species develop wing buds in later instars. The final moult to adult is around the fourth to sixth instar in grasshoppers and up to the tenth instar in crickets. The whole process takes a few weeks in good conditions and up to many months if food supply and weather are adverse.

Biology

Most of the true, short-horned grasshoppers live on vegetation near the ground and in open situations. They are active during the day, feeding mainly on grasses. Crickets and the long-horned grasshoppers (like katydids) tend to be more nocturnal and frequent all heights of vegetation, the ground and burrows. They also tend to be more omnivorous and are sometimes predatory. Often more noticeable than the insects themselves is the myriad of species-specific sounds produced mainly by the males. The largely nocturnal cricket 'songs' tend to be more complex and sustained for longer periods than the simple chirps and clicks of diurnal grasshoppers. In katydids, which often sit on the tops of trees, the song can be the only means of knowing that a particular species is in the area.

Apart from some notable exceptions like migrating locusts, orthopteroid insects are not strong

CRICKETS AND GRASSHOPPERS

Australian plague locusts laying eggs, *Chortoicetes terminifera*, **Acrididae** (35 mm long)

fliers. They use their powerful hind legs for quick getaway jumps and to launch into short flights.

Classification

There are two suborders along what is roughly the cricket–grasshopper division.
Crickets (suborder Ensifera):
- The antennae have more than 30 segments and are generally long (half to several times body length).
- The females possess an ovipositor (a tubular or sword-like projection from the end of the abdomen used for egg laying).

Grasshoppers (suborder Caelifera):
- The antennae are less than 30 segments long and generally less than half body length.
- The females do not possess an ovipositor.

This division is muddied by the use of the term 'long-horned grasshoppers' for the Tettigoniidae which are in the cricket suborder. However, their characteristics easily separate them from the grasshopper suborder as defined above. The following key separates the superfamilies of crickets.

Crickets (Suborder Ensifera)

- All tarsi are three segmented: superfamily Grylloidea.
- All tarsi are four segmented:
  - forewings, if present, are toughened, not membranous; stridulatory files on the forewings only: superfamily Tettigonioidea;
  - forewings, if present, are soft and membranous like hind wings; stridulatory files on the side of abdomen and hind leg: superfamily Gryllacridoidea.

Superfamily Grylloidea

Field Crickets (Family Gryllidae)

The largest family is the Gryllidae with 715 species of true crickets. The typical black, ground-dwelling, loud-singing

65

CRICKETS AND GRASSHOPPERS

Black field cricket, *Teleogryllus* sp., **Gryllidae** (30 mm long)

Endacusta sp., **Gryllidae** (18 mm long)

Mole cricket, *Gryllotalpa* sp., **Gryllotalpidae** (35 mm long)

crickets of the genus *Teleogryllus* belong to this family. Some species are known as house crickets as they share backyards and homes in some areas. Other species live in most Australian habitats and many sing for long periods (especially in the tropics) in a pitch audible to human ears. In rocky country, the genus *Endacusta* is common, though often cryptic.

Ant Crickets (Family Myrmecophilidae)

The 42 species of small, wingless crickets of the family Myrmecophilidae are unlikely to be encountered as they dwell inside ant nests.

Mole Crickets (Family Gryllotalpidae)

The small family Gryllotalpidae is well known for its incredibly loud call and its digging habits. It comprises 10 species, sometimes seen outside their burrows in backyards, but more often heard via the specially shaped burrow entrance which amplifies the song. They are very stout with powerful digging forelegs and are all placed in the genus *Gryllotalpa*.

Superfamily Tettigonioidea

Family Tettigoniidae

In Australia, this group includes only the Tettigoniidae which, with 900 species, is the largest of all cricket families. These are the bush crickets, katydids and longhorned grasshoppers.

The katydids (subfamily Phaneropterinae) live mainly on shrubs and trees and their varied, often sparse calls have been recorded as an identification tool. Most species are green and many are cryptic, their wings resembling leaves, like the many species of *Caedicia* and *Torbia*. Species are especially numerous in the wetter tropics. The longhorned grasshoppers (subfamily Conocephalinae) live mainly on grasses and can be easily distinguished from true

CRICKETS AND GRASSHOPPERS

Katydid, *Caedicia* sp., **Tettigoniidae** (45 mm long)

Katydid, *Torbia* sp., **Tettigoniidae** (50 mm long)

Longhorned grasshopper, *Conocephalus semivittatus*, **Tettigoniidae** (28 mm long)

grasshoppers by their very long antennae. The widespread genus *Conocephalus* has over 30 species. Most species are herbivores.

Superfamily Gryllacridoidea

This group contains the family Gryllacrididae with 125 species and three smaller families.

Family Gryllacrididae

These are mainly brown, soft-bodied crickets that can be found at night wandering on the ground or sometimes on vegetation. Many live in burrows or under logs, and a few produce silk to roll a leaf as a daytime hiding place. Most feed on seeds, leaves and detritus, although some are predatory.

The genus *Nunkeria* from north Qld is typical of the winged forms. Living in burrows in the dry interior are wingless species of the genus *Pareremus*.

King Crickets (Family Stenopelmatidae)

The 15 species of Stenopelmatidae include the king crickets. Many are very large and wingless and have huge biting mouthparts that make them fearsome to behold. They are predators restricted to east-coast forests.

67

CRICKETS AND GRASSHOPPERS

Nunkeria brachis, **Gryllacrididae** (30 mm long)

Pareremus sp., **Gryllacrididae** (25 mm long)

Camel Crickets (Family Rhaphidophoridae)
The 30 species of Rhaphidophoridae are known as camel crickets due to their body being held humped rather than straight. They are wingless and seldom seen because many are cave-dwelling scavengers.

Family Cooloolidae
The smallest family is the Cooloolidae which has three described species of strange, short-antennaed crickets which have only been recently discovered because they live almost entirely underground in sandy coastal Qld. The first described species was dubbed the 'Cooloola monster'.

Grasshoppers (Suborder Caelifera)
- All tarsi are three segmented: families Eumastacidae, Pyrgomorphidae, and Acrididae
- The tarsi of the foreleg and mid-leg are two or less segmented: families Tridactylidae, Tetrigidae and Cylindrachetidae.

King cricket, *Penalva* sp., **Stenopelmatidae** (50 mm long)

D. RENTZ

CRICKETS AND GRASSHOPPERS

Cooloola monster, *Cooloola propator*, **Cooloolidae** (25 mm long)

Mating **Eumastacidae–Morabinae** (35 mm long)

Family Eumastacidae

The family Eumastacidae includes the monkey grasshoppers. The majority of its 200 species are very elongate and thin, and all are wingless. They are well camouflaged among the grasses on which most feed. One genus differs from this form and can be a striking sight in the tropics: *Biroella* are wingless, often colourful and have the peculiar habit of sitting with their back legs rotated outwards and held flat.

Family Pyrgomorphidae

The family Pyrgomorphidae has only 24 species in Australia. In general, pyrgomorphid species differ from the acridids in having a conically pointed head (rather than the blunt-ended norm), as in the typical genus *Atractomorpha*. Many are winged and some, like the genus *Monistria*, are very boldly coloured. This serves as a warning of their possible distaste to predators, derived from a habit of eating aromatic plants.

Monkey grasshopper, *Biroella* sp., **Eumastacidae–Biroellinae** (18 mm long)

Locusts, Grasshoppers (Family Acrididae)

The family Acrididae with 712 species has six subfamilies in Australia and therefore a fair degree of variation. The most notorious and morphologically typical members are the locusts. In Australia, the most important is the plague locust *Chortoicetes terminifera* which can build up in numbers to very damaging plagues and migrate widely over central and eastern Australia. One of the largest grasshoppers is the often brightly coloured *Valanga irregularis*, which can be a pest of shrubs and trees in northern Australia.

69

CRICKETS AND GRASSHOPPERS

Monistria pustulifera **Pyrgomorphidae** (40 mm long)

Above: *Atractomorpha* sp., **Pyrgomorphidae** (22 mm long)

Above right: 4th instar nymphs of *Valanga irregularis*, **Acrididae** (35 mm long)

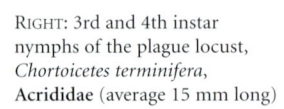

Right: 3rd and 4th instar nymphs of the plague locust, *Chortoicetes terminifera*, **Acrididae** (average 15 mm long)

CRICKETS AND GRASSHOPPERS

4th instar nymph of *Goniaea* sp., **Acrididae** (18 mm long)

Raniliella sp., **Acrididae** (30 mm long)

Sand grasshopper, *Urnisiella* sp., **Acrididae** (28 mm long)

Heteropternis sp., **Acrididae** (70 mm long)

71

ABOVE: Pygmy grasshopper, Tetrigidae (12 mm long)
RIGHT: *Tridactylus australicus*, **Tridactylidae** (4 mm long)

Not all species are restricted to a grass diet. Many feed on leaves and one large genus, *Goniaea*, feeds on dead eucalyptus leaves, with the nymphs imitating their food for concealment. In the deserts many cryptic forms abound after good rains provide rare seasons of feed. Species of *Raniliella* imitate rocks and species of *Urnisiella* hide well on sand.

Members of the large genera *Coryphistes* and *Heteropternis* spend much of their time on bark, often blending in very well despite their large size, such as the 70 mm species of *Heteropternis* from SA.

Pygmy Grasshoppers (Family Tetrigidae)

The family Tetrigidae with 70 species are known as pygmy grasshoppers. They are easy to identify by the pronotum forming a long, usually pointed shield over the rest of the body, which is often wingless. Most are less than 10 mm long and live in damp areas along creeks and ponds and in rainforests. They feed on minute plants and are capable of moving on and under water.

Family Tridactylidae

The nine species of Tridactylidae are minute (5 mm or less in length), shiny black and winged; they are also found near water. *Tridactylus australicus* is found over most of the country. Despite their size, they are probably the most powerful jumpers of all orthopterans.

Sandgropers (Family Cylindrachetidae)

The six species of the family Cylindrachetidae are superficially very similar to the mole crickets, as both live underground and are medium sized. They differ by having no wings, very short antennae, and their mid and hind legs are virtually the same short, stout shape and not adapted for jumping like most other orthopteroid insects. Apart from being accidentally dug up in sandy soil, they are rarely seen.

Further Reading

Rentz, D.C.F. (1991), 'Orthoptera' (ch. 24) in *The Insects of Australia*, CSIRO, Melbourne.

Rentz, D.C.F. (1995), *Australian Orthopteroid Insects*, University of NSW Press, Sydney.

Rentz, D.C.F. (1996), *Grasshopper Country*, University of NSW Press, Sydney.

STICK INSECTS AND LEAF INSECTS
ORDER PHASMATODEA

World families: 3; species: 2000 **Australian families: 2; species: 150**

Characteristics of Order
Incomplete life cycle; medium to very large (30–300 mm in body length); usually very elongate and cylindrical or flattened body—the majority of species resemble sticks, grasses or leaves; wings, when present, with forewings that are short and hardened into tegmina that cross over right onto left and only partly cover the membranous hind wings; long and usually thin legs, capable of being held flat along the body; short to long, filiform antennae. Nymphs usually resemble wingless adults.

Life Cycle
Sexual dimorphism is pronounced, with the males usually smaller but with larger and more functional wings. The often flightless females are found by the flying males possibly by attraction to a female scent or pheromone. Females lay 100 to 1000 eggs by dropping them to the ground singly, where they resemble seeds. The eggs hatch in the same season or up to several years later depending on conditions. Parthenogenesis (the hatching, usually of females only, from unfertilised eggs) occurs in some species where males are rare or unknown.

The nymphs resemble wingless adults in most instars. In some species, a brightly coloured first instar, sometimes mimicking ants, occurs. A few weeks to months after hatching, the nymphs undergo the final moult to adult.

Biology
The excellent camouflage of most species makes stick insects hard to see in nature, but it is also true that most species exist in low numbers. The majority live in trees and shrubs, some in grasses, and others in leaf litter. To avoid compromising their cryptic potential, most move about to feed and mate only at night. Some males fly to light. All species feed on foliage and usually eat a large variety of plant species. While their cryptic forms hide them from many predators, especially birds which hunt by a 'search image' of edible shapes, they are eaten by some birds, reptiles and mammals and by many insects. In the event of attack, some species have secondary surprise defences like a sudden opening of brightly coloured wings; or falling to the ground and not moving among the real sticks; or even hissing and kicking with strong, spiny, hind legs.

Colour change in individuals in response to environmental factors has been recorded, while some species show a great variety of colours or forms within a single population. In addition to the mostly uncommon species, three pest species are known that reach plague proportions at irregular intervals and defoliate whole forests.

STICK INSECTS AND LEAF INSECTS

Female *Eurycnema goliath*, **Phasmatidae** (200 mm long)

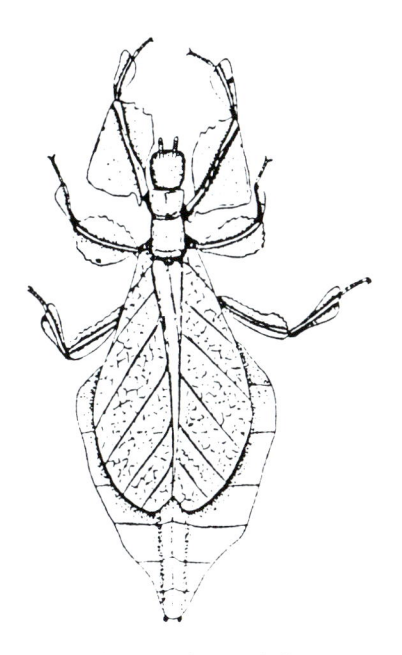

STICK INSECTS AND LEAF INSECTS

Figure 27. The true leaf insect *Phyllium* sp. [after Imms]

Female spiny leaf insect, *Extatosoma tiaratum* Phasmatidae (140 mm long)

Classification

There are only two families (Phylliidae and Phasmatidae) and, until recently, the leaf insect family Phylliidae had only one species recorded in Australia. Reclassification of parts of the Phasmatidae into this family has increased the number to about 50 species and most of these are stick-like. The true leaf insects are now a small subfamily, the Phylliinae, which is widespread in SE Asia and New Guinea, but has only three species in Australia. *Phyllium siccifolium* found in north Qld is excellently disguised and rarely seen (see Figure 27).

The family Phasmatidae with about 100 species contains the rest of this order. Notable species include the very large and showy tropical *Eurycnema goliath*, with females up to 250 mm in length; *Extatosoma tiaratum* with a very

1st stage nymph of *Extatosoma tiaratum* Phasmatidae (14 mm long)

large (150 mm), fat and spiny wingless female, a smaller, winged male and a brightly coloured ant-mimicking first instar; and *Megacrania batesii* (140 mm) from *Pandanus* plants in northeast Qld, which can be bright blue—a very rare colour in insects.

STICK INSECTS AND LEAF INSECTS

Top: *Parasipyloidea* sp., **Phasmatidae** (centre, vertical) (80 mm long)
Left: *Megacrania batesii*, **Phasmatidae** (110 mm long)

Among the more cryptic examples is the genus *Parasipyloidea*, with a wingless species living on spinifex grasses in the Australian interior.

Further Reading

Key, K.H.L. (1991), 'Phasmatodea' (ch. 25) in *The Insects of Australia*, CSIRO, Melbourne.
Rentz, D.C.F. (1995), *Australian Orthopteroid Insects*, University of NSW Press, Sydney.
Rentz, D.C.F. (1996), *Grasshopper Country*, University of NSW Press, Sydney.

WEB SPINNERS OR EMBIIDS
ORDER EMBIOPTERA

World families: 8; species: 200 **Australian families:** 3; species: 65

Characteristics of Order

Incomplete life cycle; small to medium (4–15 mm in body length); elongate sub-cylindrical body with the most obvious order characteristic being enlarged, bulbous front tarsi which house silk-producing organs; short legs; males usually have two pairs of wings with pigmented veins and/or entire wing surface; females are wingless; two short cerci at the tip of the abdomen are unusual for being asymmetrical in the males.

Life Cycle

Web spinners are gregarious with several females and their last progeny found together in silk-lined tunnelled galleries. After mating, females usually disperse to lay eggs and start new colonies. The nymphs resemble the adults, and only the males develop wings at the last moult.

Biology

The silken galleries of web spinners are constantly expanded into a system of tunnels from a hidden central area to new food sources, which include leaves, bark, mosses and lichens. As these are usually located under rocks, bark or leaf litter, embiids are very rarely encountered. Colonies also

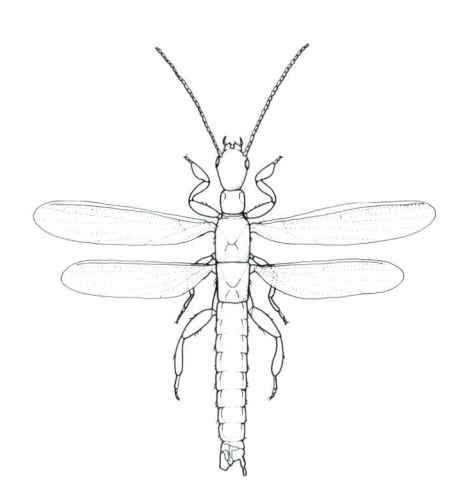

Figure 28. Winged male of *Oligotoma*, **Oligotomidae**. Note the bulbous front tarsi. [after Tillyard]

Wingless female web spinner, *Australembia* sp., **Australembiidae** (12mm long)

have escape tunnels in case of attack and embiids show an excellent ability to run backwards in the silken tubes. The females look after their eggs but the young soon disperse to extend into new colonies or face the possibility of being eaten by the female. The males are very short lived and do not feed as adults.

Classification

Three families are represented in Australia; and their total of 65 species, at first, seems like a very large proportion of the world total of 200. However, the current estimate for the world is 2000, mainly unnamed, species, with the majority found in the tropics.

The six species of the family Notoligotomidae are usually found under stones, in rock ledges and under bark in the drier parts of the continent. Their food is mainly lichens.

The 32 species of the family Australembiidae have wingless adults in both sexes and the majority are found in leaf litter along the eastern coast of the Australian mainland and in Tasmania.

The 27 species of the family Oligotomidae have a variety of habits and include the introduced genus *Oligotoma* which is one of the most likely to be encountered (see Figure 28).

Further Reading

Ross, E.S. (1963), 'The families of Australian Embioptera, with descriptions of a new family, genus and species', *Wasmann Journal of Biology*, vol. 21, pp. 121–36.

Ross, E.S. (1991), 'Embioptera' (ch. 26) in *The Insects of Australia*, CSIRO, Melbourne.

BOOKLICE AND PSOCIDS
ORDER PSOCOPTERA

World families: 35; species: 3000 **Australian families:** 26; species: 300

Characteristics of Order

Incomplete life cycle; minute to small (< 1–10 mm in length); stout bodied with a large head and laterally protruding eyes; chewing, mandibulate mouthparts; medium to long fine, filiform antennae; wings, when present, are membranous and held tent-like along the body.

There is some superficial confusion with aphids (Hemiptera), but the large head, with biting rather than sucking mouthparts, easily distinguishes the psocids.

Life Cycle

Many species engage in courtship displays before mating. Eggs are often hidden under bark, loose debris, or even silk-fastened debris.

The nymphs average six moults to the adult stage and start developing wing buds as early as the second instar.

Winged psocid, *Sigmatoneura formosa*, **Psocidae** (8 mm long)

Biology

Only a small number of species live in association with humans, feeding on stored products and cellulose-based materials like paper. The majority live on vegetation, or under bark or stones, in most parts of the continent. They feed on minute organic items such as plant spores, algae, lichens and fungi. Some species live in groups and can build up to very large numbers when a good food source becomes available.

Classification

The order is divided into three sub-orders with the majority of the species placed in the Psocomorpha which are characterised by the antennae usually being 13-segmented.

The Trogiomorpha, with antennae being up to 50-segmented, includes families with some cosmopolitan household pest species. The psocids most likely to be encountered are in the Troctomorpha. This small sub-order includes the family Liposcelidae which is characterised by enlarged femora on the hind legs and a flattened, often wingless, body. Species of the genus *Liposcelis* include the minute (1–2 mm in length) booklouse (Figure 29) and other pests of stored products and museum collections.

Further Reading

Smithers, C.N. (1972), 'The classification and phylogeny of the Psocoptera', *Memoirs of the Australian Museum*, vol. 14, pp. 1–349.

Smithers, C.N. (1991), 'Psocoptera' (ch. 28) in *The Insects of Australia*, CSIRO, Melbourne.

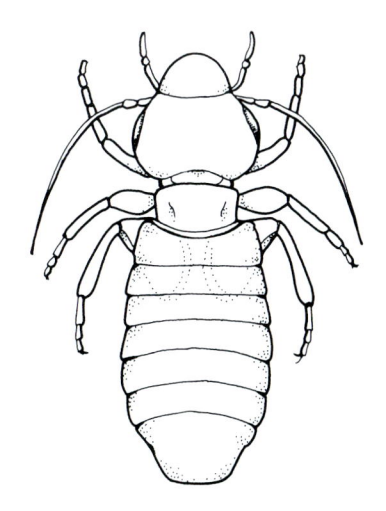

Figure 29. The booklouse *Liposcelis* sp. (family Liposcelidae), 0.8 mm. [R. Redman]

LICE
ORDER PHTHIRAPTERA

World families: 17; species: 3000 **Australian families: 14; species: 255**

Characteristics of Order

Incomplete life cycle; minute to small (0.5–10 mm in length); always dorsoventrally flattened, wingless and usually colourless ectoparasites which are found on birds and mammals; very short, three- to five-segmented antennae; eyes are reduced or absent; mouthparts are either biting mandibles as in the bird lice, or a piercing proboscis as in the sucking lice; short legs end with either single or double claws for gripping fur or feathers.

Some lice are superficially similar to some psocids (order Psocoptera), but can be distinguished best by the short stubby antennae of the lice.

Life Cycle

Lice are short lived, completing their life cycle within two weeks to two months. In some species, males are rare and females can lay unfertilised eggs which hatch into other females (parthenogenesis). Eggs are laid daily during the short adult life, with the human louse producing up to nine eggs a day. Eggs are glued to hair or fine feathers close to the skin of the host. The three to four nymphal instars look much like the adults.

Biology

All lice are ectoparasitic on mammals and birds, and most are specific to one host and sometimes to only one area of the host's body. Mammal hosts have up to three species of lice, while bird hosts can carry up to six species. Lice live their short lives hiding from light in the hair and feathers. The bird lice feed on the feathers and skin of their hosts, while most mammalian lice feed directly on the blood. A special proboscis punctures the skin and the blood is sucked in mechanically. Distribution of lice only occurs on contact between hosts such as during mating or nesting behaviour because lice cannot survive long off their hosts. Many species are potentially dangerous as carriers of blood-borne diseases, with epidemic typhus being the most important disease transmitted to humans.

Classification

This order has sometimes been divided into two suborders separating the bird lice and mammal lice. Currently, it is divided into four suborders, with only three represented in Australia. The suborder Amblycera with 107 species has both bird and marsupial parasites. The suborder Ischnocera with 123 species contains the largest family, the Philopteridae, the species of which all have bird hosts; and the Trichodectidae which has many introduced pests of farm and domestic animals.

LICE

Figure 30. Typical lice:
(a) the sheep louse, *Bovicola ovis* (family Trichodectidae);
(b) the cattle louse, *Haematopinus* sp. (family Haematopinidae).
[Roy. Zoo. Soc., NSW]

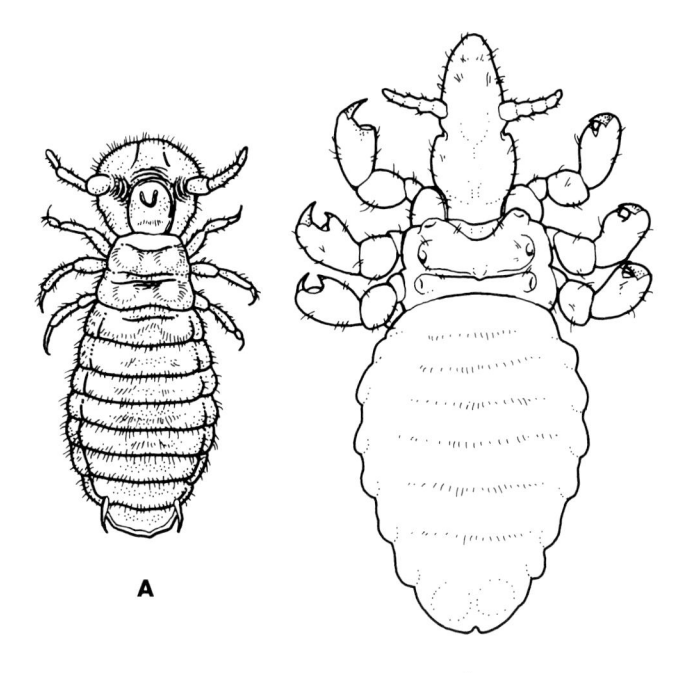

The suborder Anoplura with 25 species are all mammal specialists which even live on seals in the open ocean. There are families which specialise in cattle (Haematopinidae), dogs and rodents, and two families which attack humans. The family Pediculidae contains the head and body louse *Pediculus humanus* and the family Pthiridae has the dreaded crab louse *Pthirus pubis*. See Figure 30 for examples of typical lice.

Further Reading

Calaby, J.H. and Murray, M.D. (1991), 'Phthiraptera' (ch. 29) in *The Insects of Australia*, CSIRO, Melbourne.

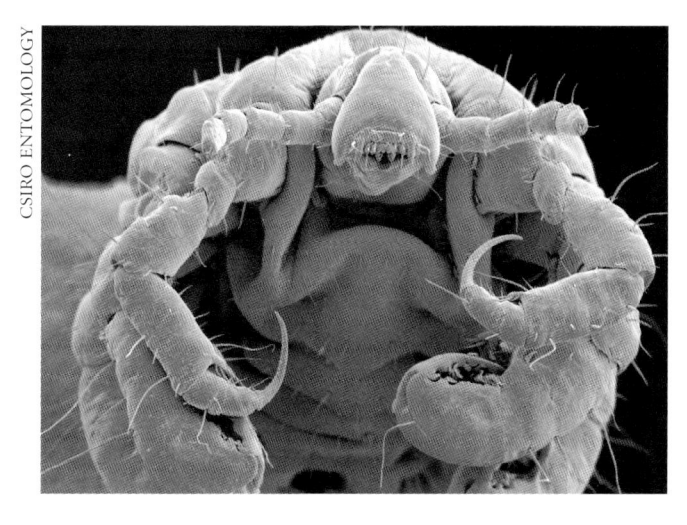

Crab louse, *Pthirus pubis*, **Pediculidae** (2 mm long)

TRUE BUGS, HOPPERS, SCALE INSECTS AND APHIDS
ORDER HEMIPTERA

World families: 133; species: 60 000 **Australian families:** 99; species: 5650

Characteristics of Order

Incomplete life cycle; minute to large (1–110 mm in length); an order with extreme diversity of general body form but all linked by the modified sucking mouthparts (rostrum), where all parts are fused into a single, often sclerotised, sharply pointed tube; antennae are commonly short with less than 10 segments. The true bugs, when winged, have forewings that are hard and shield-like at the base and membranous at the tip (hemelytron), crossing over and hiding most of the membranous hind wings, and sitting flat on the body. The hoppers have hard forewings and these are held rooflike over the membranous hind wings. The aphids and other 'soft' bugs like scale insects are often wingless or have only membranous wings and sometimes only forewings.

Immature stages are often wingless versions of adults but many more complex life cycles are found. Further diversity of body form is treated within the Classification section.

Some bugs could be superficially confused with beetles (Coleoptera) from which they mostly differ by the sucking mouth parts. Resemblance to cockroaches (Blattodea) can be settled likewise and by the cockroaches possessing long filamentous, multisegmented antennae. Some booklice (Psocoptera) can resemble aphids, but again the many jointed antennae sets the psocids apart.

Life Cycle

There are three suborders in Hemiptera: the true bugs (Heteroptera) and hoppers (Auchenorrhyncha) share many traits; but the aphid and scale insect suborder (Sternorrhyncha) is so divergent and varied it will be treated separately here and in the Biology section.

Suborders Auchenorrhyncha and Heteroptera

The plant hoppers and true bugs share a simple egg, wingless, free-living nymph instars and mostly winged adult cycle. In true bugs and most plant hoppers, the nymphs resemble the adults, although they can undergo several major colour changes through the instars. Some species in families such as the shield bugs (Pentatomidae) can be gregarious, especially in the first instar. In the so-called spittle bugs (Cercopoidea), the soft fleshy nymphs live encased in a froth produced by mixing air with their excreta. In the cicadas (Cicadidae), the nymphs burrow underground where they feed on roots for up to several years. They then emerge and fasten onto a tree trunk for the final moult to adult. All plant hopper nymphs are plant suckers, but among the true bugs there are also carnivorous and ectoparasitic species.

Suborder Sternorrhyncha

Most species are fairly sedentary, often living under waxy secretions or inside galls induced in the host plants. In these situations, eggs are usually laid within these protected spaces. All species in this suborder are plant-sap feeders.

The lerp insect nymphs (Psyllidae) form intricate covers composed of sugary excretions which can be shaped like shells or simple domes of white frass.

Some groups, including the aphids (Aphidoidea), give birth to live young and often do so without mating (parthenogenesis). Aphids have the ability to produce live female young that look like minute adults, or to mate and lay eggs that can lead to the production of more males. The eggs can also lie dormant through bad conditions. Aphids are extremely fecund, producing new young every day of their adult lives.

The scale insects (Coccoidea) have the ability to reproduce via eggs or several different types of live birth. The females are sedentary and larviform, wingless and even legless, living either in galls induced in the host plants or under waxy 'scales'. The complex nymphal development usually includes an active first instar (crawler), which disperses the species.

Biology
Suborder Auchenorrhyncha

Plant hopper adults are all plant-sap feeders, usually on flowering plants. Many are powerful jumpers when threatened, while others rely on camouflage. Some treehoppers mimic spines and nodes on branches; while leafhoppers include some very flat species that blend with the surface they sit on. The majority of hoppers are winged but usually make only short flights as part of the jumping reaction.

Suborder Heteroptera

The true bugs have a variety of feeding habits. The majority are plant-sap feeders, but there are notable exceptions such as the predatory assassin bugs (Reduviidae) which hunt insects. There are blood suckers of mammals and birds in the bedbug family Cimicidae. The majority of families are terrestrial, but a group of families known as the Gerromorpha (including the water striders) have the habit of living on the surface tension of freshwater and even in the open ocean. The Nepomorpha group of families (including the backswimmers and fishkiller bugs) live underwater, coming to the surface to breathe. Most of the species in these two groups are predators and scavengers of other insects.

Suborder Sternorrhyncha

As discussed above, many of these insects are sedentary, living in shelters. Their waxy and other secretions help to avoid desiccation and generally hide them from various predators. Many parasitic wasps and some specialist predators such as ladybird beetles do take a toll, but this can be reduced in the many species which produce honeydew. This is a sugary substance excreted as a by-product of feeding on the sap of plants. Many insects, especially ants, feed on this. In the process, the ants have set up complex 'farming' relationships with these bugs which includes protecting them from predators, moving them to fresh food sources and even taking them into the ants' nests at night.

Classification

Until recently, only two suborders divided the Hemiptera. The true bugs (Heteroptera) still stands; but the former Homoptera, which included such diverse forms as cicadas and scale insects, has been divided between the plant hoppers and the 'soft' bugs. Following is a list of characteristics for the three suborders and an introduction to the major families within each suborder.

List of Characteristics

Aphids, scale insects, mealybugs and others (suborder Sternorrhyncha):
- The body is usually soft and often hidden under waxy secretions.
- The head is deflexed down and backwards, making the rostrum appear to be emanating from between the front legs.
- The antennae are generally thread-like, with up to 16 segments and sometimes half the length of the body.
- The wings, if present, are either only forewings or if both pairs are present, with very reduced venation and held rooflike over the body.

Cicadas, treehoppers, leafhoppers and others (suborder Auchenorrhyncha):
- The body is usually elongate and sclerotised.
- The species are free living.
- The head is deflexed down and backwards, making the rostrum appear to be emanating from between the front legs.
- The antennae consist of a rounded basal segment and a minute, very thin bristle.

- The wings are usually present and consist of hardened forewings that sit rooflike over membranous hind wings.

True bugs, water striders, backswimmers and others (suborder Heteroptera):
- The body is elongate or rounded and sclerotised.
- The species are free living.
- The head usually points forward, with the rostrum clearly emanating from it.
- The antennae are very short to long, composed of obvious, simple segments (never more than five).
- The wings are usually present, with the forewings divided into a hard shieldlike base and a membranous tip, often crossing over (right on left) the membranous hind wings.

Following is an introduction to the families of Hemiptera most likely to be encountered.

Aphids, Scale Insects and Mealybugs (Suborder Sternorrhyncha)

Family Psyllidae

The Psyllidae has 330 species of so-called 'jumping plant lice' and lerp insects. The adults are usually free living and winged, and can also jump. The nymphs live in galls and other shelters on plants and often cover themselves in shields of lerp that can look like very intricate seashells or white frass. These are usually several millimetres across and hide an insect half that size. *Glycaspis*, a genus with 140 species, is often found on eucalypts and attended by ants seeking honeydew secretions.

Aphids (Family Aphididae)

The Aphididae contains 157 species of aphids. They have many body morphs. The ones most commonly encountered are wingless females found in

TRUE BUGS, HOPPERS, SCALE INSECTS AND APHIDS

large numbers near the softer growing parts of plants, best illustrated by the widespread rose aphid, *Macrosiphum*. Many are pests of agricultural and garden plants causing stunting of growth and transmitting viruses. The winged forms have very reduced venation, sometimes only one vein.

Superfamily Coccoidea

The superfamily Coccoidea contains more than 840 species in 13 families of scale- and gall-forming insects. In this group, the females are wingless, larviform and sedentary. The adult males have no wings or only one pair of wings, do not feed and live only long enough to mate. The four most distinctive families are illustrated here.

Family Eriococcidae

The Eriococcidae has 172 sometimes gall-forming species, some of which induce the host plant to produce very intricate structures within which the females live. The 39 species of the genus *Apiomorpha* make distinctive woody galls on eucalyptus leaves and stems.

Lerp-covered nymphs, *Glycaspis* sp., **Psyllidae** attended by ants (4 mm across)

Rose aphids, *Macrosiphum rosae*, **Aphididae** (4 mm long)

Galls of *Apiomorpha* sp., **Eriococcidae** (6 mm high)

85

Mango scale, *Aulacaspis tubercularis* **Diaspididae** (3 mm across)

Monophlebulus pilosior, **Margarodidae** (8 mm long)

Armoured Scales (Family Diaspididae)
The Diaspididae contains over 240 species of armoured scales. The females have no legs or antennae and live under waxy scales attached to the plant host by their rostrum. The mango scale, *Aulacaspis tubercularis*, can be seen with the free-living, 1-mm males grouped around the female scales.

Family Margarodidae
The Margarodidae, with 40 species, includes some mealybugs with large free-living females and winged males, sometimes with plumes arising from the tip of the abdomen. The cottony cushion scale *Icerya purchasi* feeds on *Acacia* in Australia, but devastated citrus in America when accidentally introduced there. The genus *Monophlebulus* contains mealybugs which live mainly on eucalypts.

Wax Scales (Family Coccidae)
The Coccidae, with 80 species, are the wax scales. Many gardens are blighted by the white soft scale, *Ceroplastes destructor*, which attacks mainly

Top: Cottony cushion scale, *Icerya purchasi*, **Margarodidae** (10 mm long)

Right: Pink wax scale, *Ceroplastes rubens*, **Coccidae** (4 mm across)

Cicada, *Abricta* sp., **Cicadidae** (30 mm long)

Cicada nymph 'shell' left by an emerged adult, **Cicadidae** (30 mm long)

citrus; and by the pink wax scale, *C. rubens* which helps to spread a destructive sooty mould.

Cicadas, Treehoppers, Leafhoppers and Others (Suborder Auchenorrhyncha)

Cicadas (Family Cicadidae)

The best known members of this suborder are cicadas. Many east coast species are very familiar to children in the summer with names like double drummer, green grocer and black prince. Females lay eggs on branches of their host trees, and the newly hatched nymphs fall to the ground and burrow with their modified front legs. These feed underground on the sap in roots for up to several years before emerging in early summer for the final moult. The adults then feed on the sap of branches in the canopy. The family Cicadidae contains 250 species found throughout Australia. Their distinctive, loud calls are produced by two vibrating drumlike membranes called timbals. These are mounted on either side of the abdomen and resonate in large cavities that amplify the sound. Each species has a distinct song that can identify it by sound alone. *Abricta* is a colourful genus from the tropics.

Leafhoppers (Family Cicadellidae)

The Cicadellidae with 664 species is the largest family of several families of

Leafhopper, *Ledromorpha planirostris*, **Cicadellidae** (16 mm long)

TRUE BUGS, HOPPERS, SCALE INSECTS AND APHIDS

leafhoppers. Most are elongate, green or brown and even bluish in colour, and under 20 mm in length. Unlike many other families of hoppers, which tend to have several well-defined spines along the hind tibia, they have one or more rows of fine hairlike spines. The largest, at 28 mm, is the very flat *Ledromorpha* species which spends much time sitting still, camouflaged on the trunks of eucalypts. A different and typical shape for leafhoppers is represented by *Brunotartessus*, which also lives on eucalypts.

Treehoppers (Family Membracidae)

The Membracidae contains 57 species of treehoppers. They are very distinctive with the pronotum enlarged into a dome over the head and a pointed extension running back along the centre of the body between the wings. This can be very ornate as in the 4-mm *Eutryonia monstrifer*.

ABOVE LEFT: Leafhopper, *Brunotartessus fulvus*, **Cicadellidae** (8 mm long)

ABOVE RIGHT: Treehopper, *Eutryonia monstrifer*, **Membracidae** (5 mm long)

LEFT: Gumtree hopper, *Eurymeloides pulchra*, **Eurymelidae** (12 mm long)

TRUE BUGS, HOPPERS, SCALE INSECTS AND APHIDS

Family Eurymelidae
The 84 species of the family Eurymelidae are squat, broad leafhoppers which are commonly attended by ants. *Eurymeloides pulchra* is common on eucalyptus saplings.

Family Fulgoridae
The Fulgoridae is a very distinctive family of leafhoppers. It is mainly tropical with large bizarre insects in other parts of the world called 'lantern flies', but with only 20 species recorded in Australia. The heads in some genera have a long, horned process pointing forward, as in the tropical *Rentinus dilatatus*.

Family Flatidae
The Flatidae contains 96, usually laterally flattened, species. From the side, these leafhoppers always have a strongly triangular appearance with the wings rising high over the rear of the body. Some green species, especially in the genus *Siphanta*, congregate on branches and look like plant spines.

Family Eurybrachyidae
The Eurybrachyidae comprises 46 species of broad, flat, large hoppers. The many species of *Platybrachys* live on eucalypts.

Family Derbidae
A small but very distinctive family is the Derbidae, with extremely

Lantern fly, *Rentinus dilatatus*, **Fulgoridae** (16 mm long)

Planthopper, *Siphanta* sp., **Flatidae** (8 mm long)

Planthopper, *Platybrachys* sp., **Eurybrachyidae** (10 mm long)

Planthopper, *Zoraida* sp., **Derbidae** (10 mm across)

long-winged species in the genus *Zoraida*. Most species feed on grasses and live mainly in the north.

True Bugs (Suborder Heteroptera)
Family Miridae
The largest family of true bugs is the Miridae. The 600 Australian species show much variation in the generally elongate, dorsally flattened, soft body form, but all share one feature. The forewing has a triangular area known as a cuneus marked off by a crease just inside the membranous tip (Figure 31). Most species are small (2 to 6 mm in length). Yellows and greens with darker markings predominate. Some species are good mimics of ants and braconid wasps.

Family **Miridae** (12 mm long)

Seed Bugs (Family Lygaeidae)
The Lygaeidae is a large family with 401 species of similar shape to mirids, differing by not having a cuneus, a reduced or missing membranous tip of the forewing, and a more robust body. Most species are known as seed bugs. Lygaeids are mainly brown to red in colour and have two ocelli which the mirids do not have. Lygaeids feed on seeds or plant sap, and some are predators. Among the lygaeids living

Figure 31. The forewing (hemelytron) of the family Miridae, showing the pronounced triangular area, the cuneus. The shaded area is the toughened part; and the clear area, the membranous part. [R. Redman]

Family **Lygaeidae** (5 mm long)

Physopelta australis, **Largidae** (14 mm long)

on vegetation are species which congregate around the succulent growing tips. A smaller, related family is the Largidae which has the bright red and black, apparently distasteful species in the genus *Physopelta* found all along the east coast of Australia.

Assassin Bugs (Family Reduviidae)

The assassin bugs, family Reduviidae, are well-known members of this order. There are 300 species characterised by having a distinct neck between the pronotum and head, and a very visible rostrum which curves back without being held against the body like in most true bugs. All species are predators. A typical widespread and diurnal genus is *Pristhesancus*. A bite from these insects can be very painful.

Assassin bug, *Pristhesancus* sp., **Reduviidae** (20 mm long)

Crusader Bugs (Family Coreidae)

Members of the family Coreidae, with 57 species, are large (20–30 mm in length), elongate and stout, with a segmented, raised margin around the abdomen which, from above, makes the wings appear to be sitting in a depression. In many species, the males have enlarged rear legs, as in the

Crusader bug, *Mictis profana*, **Coreidae** (25 mm long)

TRUE BUGS, HOPPERS, SCALE INSECTS AND APHIDS

Podsucking bug, *Riptortus serripes*, **Alydidae** (14 mm long)

Bed Bugs (Family Cimicidae)

The Cimicidae, with only one species in Australia, deserves mention among the major groups covered here as it is the wingless, flat and worldwide bed bug, *Cimex lectularius*. In other parts of the world, different species suck the blood of other mammals, especially bats.

Superfamily Pentatomoidea

Perhaps the most typical of the true bugs are the 550 members of the shield bug superfamily, the Pentatomoidea, which has nine families. The main characteristic of these bugs is the very large scutellum (the triangle below the pronotum, between the wings) which usually extends halfway down the abdomen.

well-known and widely distributed crusader bug, *Mictis profana*.

Family Alydidae

The small family Alydidae has very elongate species that, from above, are constricted in the middle and have very thin, long legs and antennae. The podsucking bug, *Riptortus serripes*, once fed mainly on *Acacia* but is now becoming a pest of beans.

Shield Bugs (Family Pentatomidae)

The dominant and typical family is Pentatomidae, with 391 species, many

Bed bugs, *Cimex lectularius*, **Cimicidae** (4 mm long)

Mating shield bugs, *Poecilometis* sp., **Pentatomidae** (24 mm long)

Burrowing bug, *Adrisa* sp., **Cydnidae** (8 mm long)

of which are known as stink bugs and produce foul-smelling defensive liquids. The body is rounded when viewed from above, and the pronotum often extends out to the sides; unlike most other families in this group, they have five-segmented antennae. The genus *Poecilometis* has 40 species which are usually found on bark. Some pentatomids are pests of crops; but among the many plant feeders, there are also predatory species which feed on garden pests.

Burrowing Bugs (Family Cydnidae)

The Cydnidae, known as burrowing bugs, comprise 43 species of mainly shiny black bugs that can be easily identified by having spines along the tibiae of all legs. They feed mainly on roots and seeds. Sometimes after rains, they emerge in extreme numbers and inundate lights at night.

Jewel Bugs (Family Scutelleridae)

Some very distinctive insects belong to the 26 species of the jewel bug family Scutelleridae. The scutellum of this family is so large that it usually entirely covers the wings, giving the appearance of a single dome. Most species have bright metallic colours, sometimes in a variety of patterns such as on the cotton harlequin bug *Tectocoris diophthalmus* from Qld.

Cotton harlequin bugs, *Tectocoris diophthalmus*, **Scutelleridae** (16 mm long)

Family Tessaratomidae

The Tessaratomidae is a small family of large distinctive species with very small heads compared to the other pentatomoid families. The *Lyramorpha* species is 25 mm long and lives in the rainforests of Qld.

The Aquatic Bugs

There are two groups of aquatic families: the Gerromorpha which live on the surface tension of water; and the Nepomorpha which live below the surface.

Lyramorpha parens, **Tessaratomidae** (30 mm long)

The Gerromorpha

This group includes 93 Australian species in six families.

Water Striders (Family Gerridae)

The most noticed surface dwellers are the water striders with 32 species. They have very long, slender legs and move gracefully, often very quickly, over usually calm water bodies. Gerrids are hunters and scavengers, finding prey by the ripples produced by insects falling onto water. Most live on fresh water, but there are species in estuaries, and even one species which lives out in the open ocean. One of the larger Australian species is the 25-mm long *Limnometra cursitans* from northern Australia.

Water Measurers (Family Hydrometridae)

This is a small family of surface dwellers called water measurers. They are very elongate and thin, with the head three times as long as it is wide. The eight species are hard to spot as they tend to move very slowly in search of prey, usually close to the edge of ponds. A good example is the 10-mm species of *Hydrometra*.

Family Velliidae

The 39 species of this family share the habits of the water striders, but are much smaller, less than 5 mm in length, have squatter bodies and relatively short legs. Some congregate into skating groups, keeping a steady position in clear flowing streams, and

Water strider, *Limnometra cursitans*, **Gerridae** (25mm long)

Water measurer, *Hydrometra* sp., **Hydrometridae** (10 mm long)

many have a velvety sheen to their dark bodies, such as the species of the largest genus *Microvelia*.

The Nepomorpha

This group includes 127 Australian species in eight families. The species of this group have minute antennae which in most cannot be seen from above.

Backswimmers (Family Notonectidae)

Perhaps the best known water bugs are the backswimmers. The 40 species in the family Notonectidae are fast-swimming predators with raptorial front legs and long hind swimming legs. Their common name comes from the habit of swimming upside down, often near the surface, to which they periodically come for air.

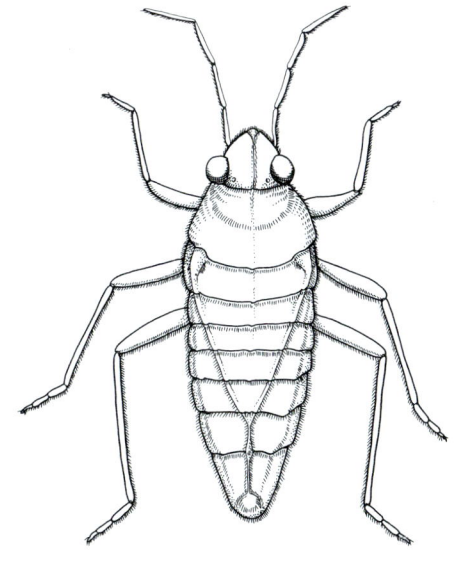

Figure 32. *Microvelia* sp., (family Veliidae), 3 mm long. [R. Redman]

TRUE BUGS, HOPPERS, SCALE INSECTS AND APHIDS

Backswimmer, **Notonectidae** (12 mm long)

Water Boatmen (Family Corixidae)

The water boatmen (31 species) are superficially similar to notonectids. However, they swim right way up, have short forelegs and long middle and hind legs, and a flatter body shape compared with the very convex backswimmers. They tend to feed on or near the bottom and so are usually only seen as they come up for air.

Water boatman, **Corixidae** (8 mm long)

ABOVE: Fishkiller bug, *Lethocerus* sp., **Belostomatidae** (70 mm long)

BELOW: Water scorpion, *Ranatra* sp., **Nepidae** (38 mm long)

Families Nepidae and Belostomatidae

The water scorpions and the fishkiller bugs or giant water bugs are distinguished by their long tail-like siphons, which they protrude through the water surface to breathe, and by their large raptorial forelegs. The water scorpions (10 species in the family Nepidae) are very slender and long legged and measure up to 40 mm (plus siphon). The fishkiller bugs (four species in the family Belostomatidae) are wide, stouter versions that can attain 70 mm in length (plus siphon). Both families hunt tadpoles, insects and small fish.

Further Reading

Carver, M., Gross, G.F. and Woodward, T.E. (1991), 'Hemiptera' (ch. 30) in *The Insects of Australia*, CSIRO, Melbourne.

Miller, N.C.E. (1956), *The Biology of the Heteroptera*, Leonard Hill, London.

THRIPS
ORDER THYSANOPTERA

World families: 8; species: 4500 **Australian families:** 4; species: 420

Characteristics of Order

Incomplete life cycle; minute to small (0.5–15 mm in length); cylindrical elongate body; mouthparts are modified into a complex set of stylets which can both rasp and suck; legs do not end in tarsal claws but in a unique bladder-like organ; two pairs of wings, when present, are slender, membranous and with a fringe of long hairs.

Despite their small size, thrips are well known due to a number of mainly introduced species which are pests in gardens and crops. However, the majority of native species are not injurious and lead a variety of largely hidden lives.

Life Cycle

Thrips can be male or female, or bisexual. In some species, males are rare to unknown, and therefore parthenogenesis (the production of female offspring from unfertilised eggs) is commoner than sexual reproduction. Eggs are either laid on plants or inserted into plant tissue. The young are similar to the adults but lack wings. They undergo around four moults followed by a prepupal and a pupal stage. The pupal stage of thrips is also similar to the adult stage in appearance and can move if disturbed.

Biology

The majority of thrips species are found in eastern Australia with virtually no species occurring in the deserts or Western Australia.

Feeding habits are varied. The primitive species are generally fungus feeders. Many others are pollen and flower feeders, spending their lives inside the flowers. The species of economic importance are leaf and sap feeders. The important pests which damage leaves and/or fruit of plants like citrus, pears, apples, onions, bananas, orchids and garden flowers

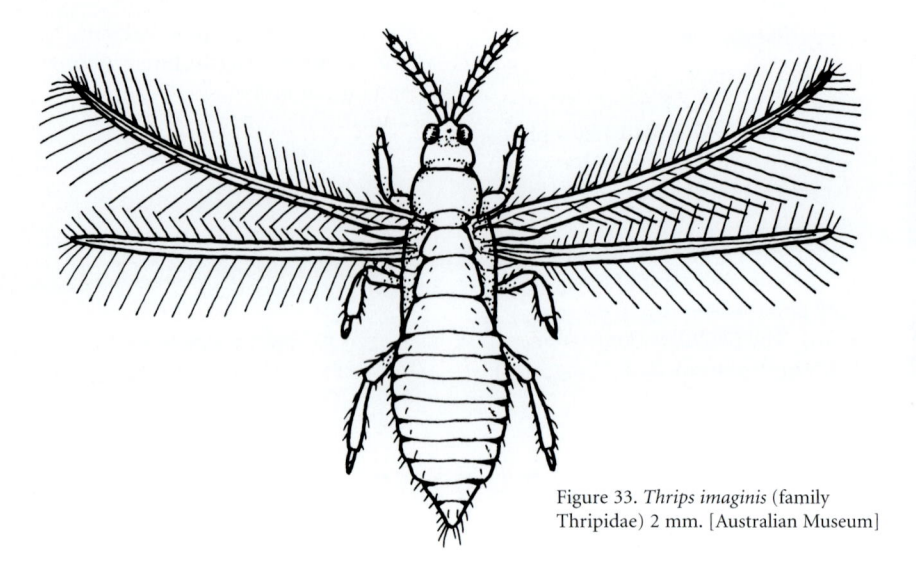

Figure 33. *Thrips imaginis* (family Thripidae) 2 mm. [Australian Museum]

are almost all introduced species. Despite the pest species, there are also predatory thrips, some of which are important in controlling other pests like scale insects.

Some species produce poisonous chemicals to deter predators; but most can and do fall prey to a variety of insects, in particular lacewing larvae, flower bugs and predatory and parasitic wasps.

fungus, insect and flower-feeding species. Many of the leaf feeders cause the formation of galls which, in some cases, enclose the thrips population completely from the outside world.

Classification

The Australian fauna fits mainly in two of the four families present.

Family Thripidae

The family Thripidae, with 128 species, contains most of the leaf and flower pest species including the greenhouse thrips (*Heliothrips* sp.), the redbanded thrips (*Selenothrips* sp.), and the plague thrips (*Thrips imaginis*) which attacks apples and pears (see Figure 33).

Family Phlaeothripidae

The family Phlaeothripidae, with 267 named species, is the largest and most varied family which includes leaf,

Giant thrip, *Mecynothrips* sp., **Phlaeothripidae** (6 mm long)

These species feed on native plants only, including *Acacia* and *Melaleuca*. The family includes one of the largest thrips in the world, *Idolothrips marginata*. It is up to 7 mm in length and feeds on fungal spores on eucalyptus leaves.

Families Aeolothripidae and Merothripidae

The small Aeolothripidae are predatory on other arthropods found mainly in flowers; and the three species of the Merothripidae are fungus feeders living in leaf litter.

Further Reading

Lewis, T. (1973), *Thrips, Their Biology, Ecology and Economic Importance*, Academic Press, London.

Mound, L.A. and Heming, B.S. (1991), 'Thysanoptera' (ch. 31) in *The Insects of Australia*, CSIRO, Melbourne.

ALDERFLIES AND DOBSONFLIES
ORDER MEGALOPTERA

World families: 2; species: 300 **Australian families**: 2; species: 26

Characteristics of Order

Complete life cycle; medium-sized (wingspan: 20–100 mm); mandibulate mouthparts; long tapering, filiform antennae; two pairs of large membranous, similar sized wings, which often have markings; slender legs with five-segmented tarsi; larvae are fully aquatic predators with strong mandibles, clawed legs and tracheal gill filaments along the abdomen.

At first glance, it is possible to confuse this order with stoneflies (Plecoptera). The most visible external characteristics which differentiate Plecoptera from Megaloptera are hind wings that are broader and larger than forewings; tarsi in adults are three segmented; the abdomen in larvae and adults usually ends in two long filaments (cerci).

Life Cycle

All species have aquatic larvae. The adults live near streams where they mate and females lay 200 to 3000 eggs close to, but out of the water. These hatch after a few weeks and the larvae then enter the water where they live as predators, moulting through up to 12 instars before pupating. The pupal stage is only known for a few species and in these, the larvae move into the soil or litter to pupate. The adults emerge several weeks later. In the tropics, a generation takes about one year, but in cold climates may take up to five years.

Biology

Most species live in clear streams as larvae, and are restricted to eastern Australia from Cape York to Tasmania. Only one species of alderfly is found in the southwest of WA. While the general world trend is for

Dobsonfly, *Archichauliodes* sp., **Corydalidae** (50 mm long)

greatest speciation in the cooler regions, in Australia this trend is reversed with over half the species being found in tropical Qld.

The larvae are predators of other aquatic invertebrates and are in turn an important food item for other predators. In other parts of the world they play an important role in the life cycle of trout and are known to anglers as a favourite bait.

The adults tend to live near the water and few are strong fliers. Some are active in daylight but most fly mainly in the evening and at night, when they will come to lights. They feed very little if at all in the short-lived adult phase.

Classification

In the past, this order has been treated as a group within the lacewing order Neuroptera. While there are many similarities in adults, an important larval difference is that Neuroptera species have modified sucking mouthparts instead of the chewing mandibles of Megaloptera.

The two Australian families are the alderflies (Sialidae) and dobsonflies (Corydalidae), which are relatively easy to differentiate. Sialid adults have no ocelli and are small with wingspans less than 30 mm; while corydalids have three ocelli and wingspans greater than 30 mm. The sialid larvae have seven pairs of gill filaments and

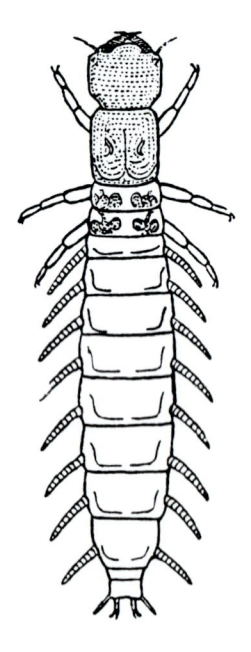

Figure 34. Larva of *Archichauliodes*, Corydalidae, 27 mm. [R.J. Tillyard]

an extra pointed filament at the tip of the abdomen; while the corydalids have eight pairs of gills and no extra filament (see Figure 34).

Family Sialidae

The four known species of alderflies are restricted to the east coast from Tasmania to Cape York. They are rarely seen, though the Tasmanian *Austrosialis ignicollis* is visually striking with an orange prothorax against a black body.

Family Corydalidae

Twenty of the 23 named species of dobsonflies are in the one genus, *Archichauliodes*, which is found mainly in eastern Australia, with one species isolated in WA.

Further Reading

Theischinger, G. (1983), 'The adults of the Australian Megaloptera', *Aquatic Insects*, vol. 5, pp. 77–98.

Theischinger, G. (1991), 'Megaloptera' (ch. 32) in *The Insects of Australia*, CSIRO, Melbourne.

LACEWINGS, ANTLIONS, MANTIS FLIES
ORDER NEUROPTERA

World families: 19; species: 5000 Australian families: 14; species: 623

Characteristics of Order

Complete life cycle; small to large (wingspan: 5–150 mm); body elongate and soft; two pairs of generally equal-sized, membranous wings with complex venation—main veins are forked along the wing tip, sometimes with hairs along the veins and fringing the perimeter; wings are usually held over the body in a tent-like posture; mandibulate mouthparts; eyes are sometimes colourful; generally filiform antennae, which are clubbed in some families; legs have five-segmented tarsi; most larvae are active predators with the mandibles modified into long piercing and sucking antler-like tubes.

Superficially, confusion with several other orders is possible. The alderflies and dobsonflies (Megaloptera) used to be placed in this order. This order differs as larvae by having chewing mandibulate mouthparts, and as adults by the veins of the wings not being strongly forked at the margins. The stoneflies (Plecoptera) differ by the presence of two filaments (cerci) at the tip of the abdomen, and by holding their wings either flat or actually curved around the body. The caddisflies (Trichoptera) differ by having almost no cross (vertical) veins in their wings, which are also more hairy.

Life Cycle

In about half the families, eggs are laid on thin stalks, either in rows or in a 'U'-shaped cluster, attached to wood or leaves. Others lay directly onto the substrate, singly or in rows. Most families have active, long-legged predacious larvae, the particular habits of which are discussed under their family headings (see Figure 35). Exceptions are the parasitic forms of the family

LACEWINGS, ANTLIONS, MANTIS FLIES

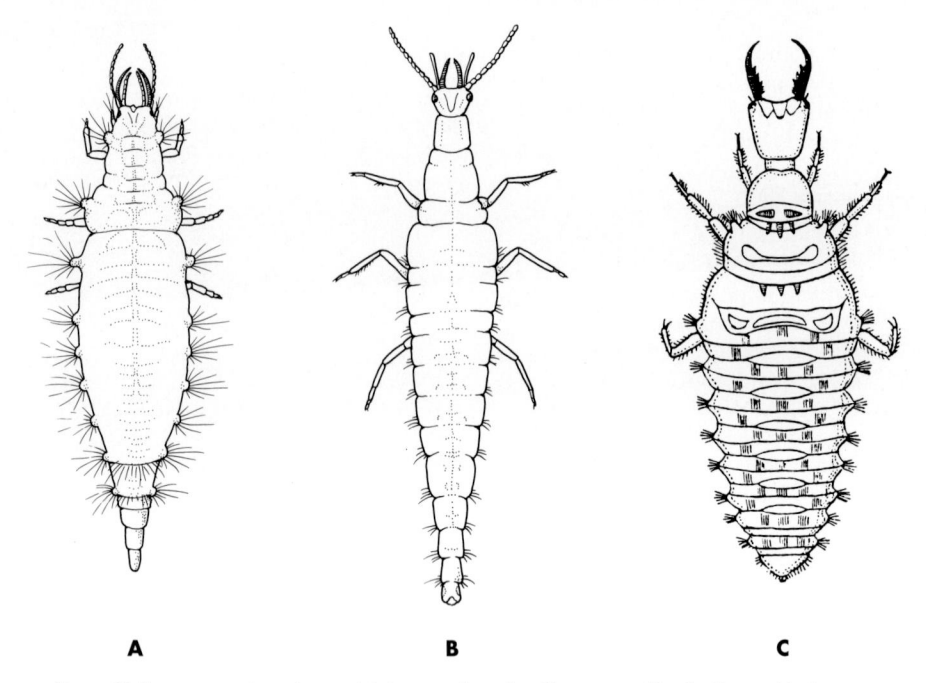

A   **B**   **C**

Figure 35. Some neuropteran larvae: (a) the green lacewing *Chrysopa* sp. (family Chrysopidae), 20 mm; (b) the brown lacewing *Micromus* sp. (family Hemerobiidae), 10 mm; (c) the antlion *Acanthaclisis* sp. (family Myrmeleontidae), 25 mm. [after Dept. of Agric. NSW & A.Tonnoir]

Mantispidae, and the burrowing detritus-feeding, grub-like larvae of the Ithonidae.

After only three, rarely four or five, moults, the larvae spin simple silken pupation chambers which are sometimes covered by debris for camouflage. On average, neuropterans have several generations per year, but some alpine species take up to two years for one generation.

Biology

Neuropterans are found in most habitats. While most larvae are predacious, adults may be predacious or omnivorous, with some lacewings feeding on honeydew or pollen. The antlions inhabit drier sandy areas allowing the formation of larval sand-pit traps. Hence, many species live in the arid interior, or in the sandstone, dry sclerophyll forests closer to the coast.

Neuropterans are prey for many birds, bats and arthropods. Some lacewings have foul defensive odours as adults, and many larvae cover themselves in debris and the bodies of their prey to disguise their edible form. As the favourite food of lacewings includes soft sap-sucking pests such as aphids and scale insects, their presence has a major biological control benefit in gardens and farms.

Classification

There is a very high level of endemism, with over 90% of genera and species restricted to Australia. By far the largest of the 14 families is the antlions, Myrmeleontidae, with about 40% of species.

Relatively simple features separate the major families. Without resorting to complex analysis of wing venation, the following sets of features will

Mantis fly, *Campion* sp. **Mantispidae** (20 mm long)

determine the correct family most of the time. Some of the smaller, more obscure families are not discussed here.

Mantis Flies (Family Mantispidae)

The easiest neuropteran family to recognise is the Mantispidae. They have the typical lacewing body with clear membranous wings, but are the only ones to have raptorial forelegs like in praying mantids. Any confusion between these two orders can be dispelled by observing that the true mantids either have no wings (in the nymph stages and some adults), or when wings are present, that the forewing is hardened and coloured, with only the hind wing being membranous. The wingspan of the 45 species ranges in size from 10 to 55 mm. They are active predators as adults, catching prey with their spiny front legs. Another interesting feature of this family is the parasitic larvae: some parasitise the egg sacs of spiders, while others use the larvae of social wasps.

NB: *The following two families are characterised by their wings often being scaled or hairy.*

Moth Lacewings (Family Ithonidae)

Major differentiating features:
- The wings are hairy along the veins and margins.
- They are mainly brown and grey.
- They appear very moth-like.

There are 14 species in a mainly Australian family. The wingspan ranges from 50 to 65 mm. Larvae differ from the normal long-legged predators, being grub-like with short legs and jaws, and burrowing in search of other insect larvae and plant matter in a life cycle of up to two years. Adults of some species swarm and come to lights where they can be quickly distinguished from moths by lacking the curled proboscis mouthparts. Most species are restricted to eastern Australia.

Dusty Wings (Family Coniopterygidae)

Major differentiating features:
- The wings have powdery white scales.

103

LACEWINGS, ANTLIONS, MANTIS FLIES

Sand traps of antlion nymphs, **Myrmeleontidae** (pits average 50 mm across)

Antlion adult, *Heoclisis* sp., **Myrmeleontidae** (40 mm long)

- They are small with a wingspan less than 15 mm.

The 50 species live on trees and shrubs and have a short-lived larval stage. Almost nothing is known about the biology of this family whose members resemble the bug family Aleyrodidae (whitefly), from which they differ by not having the sticking 'beak' mouthparts of all true bugs. They are found over most of Australia.

NB: *The following families have wings generally free of scales and if hairy, only a fine subtle layer.*

Threadwings (Family Nemopleridae)
Major differentiating features:
- The hind wing is very long and thin, and is sometimes curled and ribbon-like.

There are ten species of antlions known as threadwings. Most live in the dry interior and in WA, where their larvae are predators in dry dusty overhangs.

Antlions (Family Myrmeleontidae)
Major differentiating features:
- The antennae are short (from less than half wing length to head width), and thicken gradually towards the end (not clubbed).
- The body is long and thin, and both pairs of wings are elongate, of similar size, and often speckled.

There are 250 species with wingspans from 20 to 150 mm. The best known are the true antlions, the nymphs of which dig conical pits in dry sand to

LACEWINGS, ANTLIONS, MANTIS FLIES

capture ants and other insects. Other species are burrowing or free living on vegetation or on the ground. They are distributed all over Australia, though with more species in drier areas than on the coast. The genus *Heoclisis* has the largest species.

Owlflies (Family Ascalaphidae)
Major differentiating features:
- The antennae are prominently clubbed, and are half or more the length of the forewing.

There are 40 species which are known as owlflies, with wingspans from 30 to 100 mm. These handsome insects have a curious stance where the body is extended above the branch they are sitting on, while the wings hang downwards as shown by *Suphalomitus*. They are diurnal with adults hunting insects in flight. The very large-jawed larvae hunt prey on the ground or among low vegetation.

Green Lacewings (Family Chrysopidae)
Major differentiating features:
- The antennae are thin, filiform, tapering and are longer than half the length of the forewing but less than the whole wing.
- The wings have more vertical veins than longitudinal ones, giving the

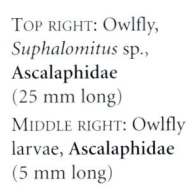

Top right: Owlfly, *Suphalomitus* sp., **Ascalaphidae** (25 mm long)

Middle right: Owlfly larvae, **Ascalaphidae** (5 mm long)

Right: Green lacewing, *Italochrysa insignis*, **Chrysopidae** (20 mm long)

Green lacewing nymph, **Chrysopidae** (13 mm long), with prey

appearance of numerous, almost vertical, rectangular cells rather than horizontal, rectangular cells, which are clear in most species.
- Body colour is usually green, with metallic-coloured eyes.

The green lacewings or 'golden eyes', with 55 species, have a wingspan from 20 to 65 mm. Adults often come to lights. The larvae are elongate and roam over vegetation, hunting soft-bodied insects. Some larvae have long hairs on to which they attach debris and bodies of their prey to provide camouflage.

NB: *Three other lacewing families occur in Australia; their differences from the chrysopids are listed below.*

Family Osmylidae
- The antennae are filiform and less than half the forewing length.
- The forewings are very broad, patterned with generally brown or grey markings and the venation is busy with small cells.

Another family character is the presence of three ocelli, but these can be hard to distinguish in the field. There are 40 species in the Osmylidae. This family is most diverse in the high country of eastern Australia. Many species are associated with water, with some larvae being aquatic predators. Wingspans range from 25 to 65 mm.

Brown Lacewings (Family Hemerobiidae)
- The antennae are moniliform, usually as long as the forewing.
- The wingspans are smaller, ranging from 10 to 25 mm, with a fine subtle layer of hair in most and more elongate than square venation cells.
- They are generally brown and other non-green hues, and the eyes are not metallic coloured.

LACEWINGS, ANTLIONS, MANTIS FLIES

Brown lacewing,
Micromus tasmaniae,
Hemerobiidae
(10 mm long)

There are 35 species of brown lacewings in the Hemerobiidae. The larvae are predators of aphids and other soft bugs. The adults live on bushes and trees, and often come to lights. *Micromus tasmaniae* is widespread.

Silky Lacewings (Family Psychopsidae)
- Both pairs of wings are very broad, with a silky sheen and fold tent-like over the body at rest.
- The wingspan ranges from 25 to 65 mm.
- The wings often have striking metallic patterns that, at first glance, make them look moth- or butterfly-like.

These are the distinctive but elusive silky lacewings, with 13 species. All but two species are restricted to the warmer parts of the eastern coast of Australia, and only a few are at all common.

Family Nymphidae
This family needs mention as it exhibits traits from several of the families defined above. The Nymphidae is endemic to Australasia with 23 large species (wingspan: 35–90 mm), with two general body types. Some, like *Myiodactylus*, look

Myiodactylus sp., **Nymphidae** (16 mm long)

like chrysopids but differ by having broad and slightly hairy wings and moniliform antennae. Others have narrow wings and look like antlions but differ by having longer antennae that are not thickened at the ends.

Further Reading

New, T.R. (1986), 'A review of the biology of the Neuroptera, Planipennia, *Neuroptera International*, suppl. 1, pp. 1–57.

New, T.R. (1991), 'Neuroptera' (ch. 34) in *The Insects of Australia*, CSIRO, Melbourne.

BEETLES
ORDER COLEOPTERA

World families: 160; species: > 300 000
Australian families: 117; species: > 28 200

Characteristics of Order

Complete life cycle; minute to large (0.4–80 mm in length); usually chewing, mandibulate mouthparts; forewings are modified into hard protective elytra which do not beat in flight, and which are held at rest longitudinally along the abdomen with the edges forming a straight line along the centre; hind wings, when present, are membranous and folded under elytra; the prothorax is visible from above as a single plate called the pronotum; legs are held below the body; larvae have a sclerotised head capsule, and their mandibles, eyes and antennae are visible while legs may or may not be present.

Beetles can be distinguished from cockroaches (Blattodea) by their forewings which are modified into elytra and meet in a straight line along the centre of the abdomen; and from true bugs (Hemiptera) by their chewing rather than sucking mouthparts; and from both orders by being endopterygotes.

Life Cycle

Many species are sexually dimorphic. The eggs are small and simple; from as few as four to many hundreds of eggs are laid per female, and are usually laid on or near the food supply. Larvae usually fend for themselves but subsocial behaviour is known. In some species, both male and female cooperate to provide the larvae with either plant material, dung or carrion. The larvae develop through several instars, their growth lasting from a few weeks to several years depending on species, biology and climatic conditions (see Figure 36). Some species have active, early instars for dispersal, followed by slow inactive later instars with a great variety of shapes. Pupation is usually not in a spun or built chamber but located either within the host plant or in the soil. Adult emergence is governed by climatic factors including temperature changes and rainfall.

Biology

As can be expected in such a huge order, the range of habits of beetles is immense. Crowson (1981) gives a good overview of the biology of world species. Lawrence and Britton (1991, 1994), Matthews (1980–2002) and Moore 1980–92) discuss Australian species.

Not all adult beetles feed, but those that do have a variety of diets and include herbivores, predators and scavengers. Herbivorous beetles feed mostly on plant foliage as adults (e.g. Scarabaeidae, Chrysomelidae, Curculionidae). Predatory species generally have longer legs and are very active (e.g. Carabidae, Cleridae), though short-legged species are found in groups with slow prey (e.g. Coccinellidae, Histeridae).

Scavengers exploit nearly all possibilities: carrion (Trogidae, Cleridae, Dermestidae); dung (Scarabaeidae, Hydrophilidae, Histeridae); leaf litter

A **B** **C** **D**

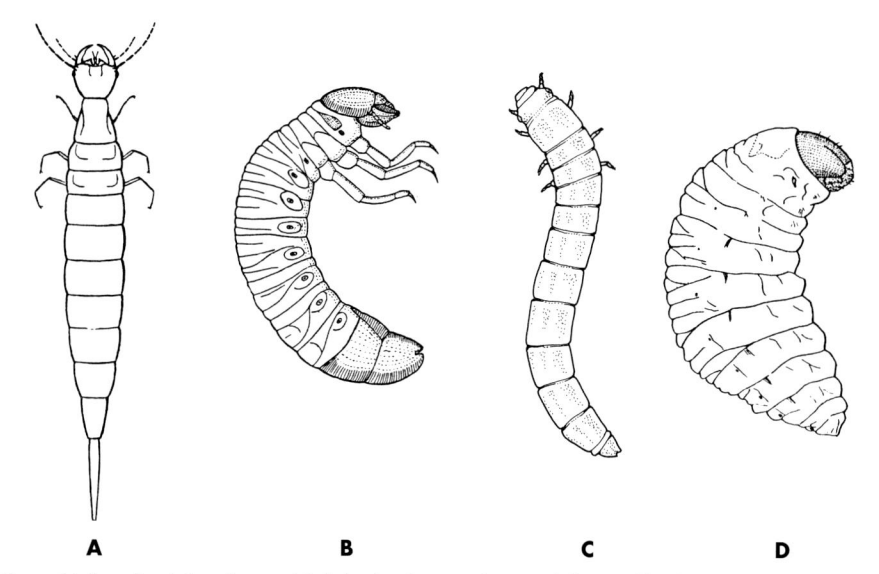

Figure 36. Some beetle larval types: (a) diving beetle, an active aquatic hunter (family Dytiscidae); (b) a 'C' grub, typical larva of family Scarabaeidae (like Christmas beetles), which lives underground; (c) a sclerotised type, typical of some Tenebrionidae and Elateridae, which lives in rotting wood; (d) the legless larva of a weevil (family Curculionidae) which burrows in wood. [R.J. Tillyard]

(Staphylinoidea); rotting wood (Cerambycidae) where they feed either directly on the substrate or on other insects (like fly larvae) that share the food source.

Larval stages have an even broader range of food sources. They feed either internally or externally on all parts of plants. Many eat fungi and some scavengers need their food source to be modified by fungi. These sometimes possess special organs for transporting fungal spores to a new host.

Long-legged predatory larvae are generally very active (Carabidae, Histeridae, Coccinellidae), while some use ambush from tunnels (tiger beetles—Cicindelinae) and others follow the tunnels of wood-boring insects in search of prey. Among the scavengers are several hundred species of dung feeders; while carrion, leaf litter and decaying vegetation are also common larval habitats.

There are 16 families of aquatic or semi-aquatic beetles. The diving beetles (Dytiscidae) trap an air bubble between the elytra and abdomen, while the Hydrophilidae have specialised hairs (plastron) on the body surface which retain large amounts of air. Aquatic larvae may or may not have various types of gills and some can puncture plant stems to obtain an air supply.

Other habitats include living in birds' nests or animal dens. Some families have associations with social insects. These mainly ant-nest intruders, known as inquilines, may feed on ant secretions, steal ant food, or feed on the ants themselves (Pselaphidae, Anobiidae, Histeridae and others). They are often highly adapted in shape and possess special hairs called trichomes which produce secretions that appease and feed the ants. Termitophiles live in termite nests and include many genera of Staphylinidae.

Classification

Family level classification is still in a state of flux. Of the 117 families recognised in Australia, three small families and numerous subfamilies are endemic. Ten families have over 800 species and the largest, the weevils (Curculionidae), has over 8000. Some southern species have Gondwanan links, with related species in New Zealand and southern South America, while some northern species have links with Asian and New Guinea faunas.

The order is divided into four suborders. The Archostemata, with only nine species and the Myxophaga with two species (which are less than 1 mm long) will not be described in this guide. The two other suborders divide the largely predatory ground- and water-dwelling Adephaga (2730 species in seven families) from the majority (> 90%) of beetles in the Polyphaga (25 450 species in 107 families).

The following characteristics help to define these two suborders (see Figure 37).

Suborder Adephaga:
* Notopleural sutures are present along the prothorax.
* The hind coxae are immovable and usually divide the first abdominal segment.
* The antennae are never with a club.

Suborder Polyphaga:
* No notopleural sutures are present along the prothorax.
* The hind coxae are movable and usually do not divide the first abdominal segment.
* Many types of antennae.

Suborder Adephaga

Ground Beetles (Family Carabidae)

This is one of the major families of beetles (1–60 mm in body length) with about 1800 species in 10 subfamilies and 279 genera.

Distinguishing features of carabids include a stout, usually flattened shape, often with striated elytra; long legs for fast movement; filiform antennae; six visible abdominal segments, the first is divided by the hind coxae.

All species are terrestrial, and all but a few are predacious as both adults and larvae. Most species live on the ground, some are associated with the edges of creeks and ponds, and others live on tree trunks and foliage. Many ground species are flightless. Carabids live in nearly all climates from hot dry deserts to cold wet mountain areas. The largest species is the 75-mm *Mecynognathus dameli* which is a ground hunter in Cape York Peninsula. The most distinctive subfamily is Cicindelinae, the tiger beetles, which can run down even flying prey. Many species of the largest

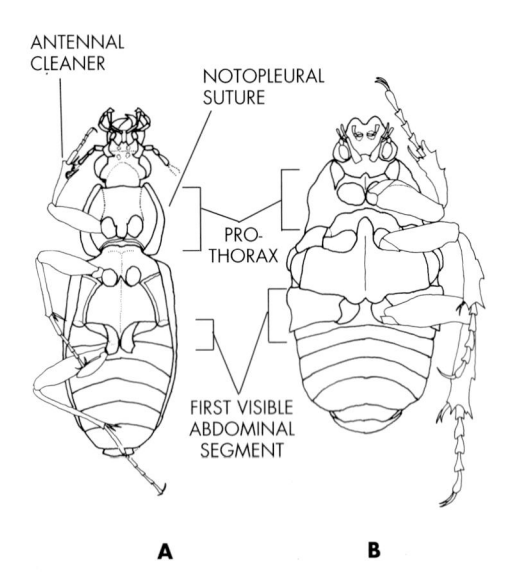

Figure 37. The underside of two beetles: (a) belongs to the suborder Adephaga and shows the notopleural suture, or groove, along the bottom inner edge of the prothorax and a divided first abdominal segment; (b) belongs to the suborder Polyphaga and shows no such suture and an undivided first abdominal segment. [R. Storey]

Ground beetle, *Mecynognathus dameli*, **Carabidae** (up to 75 mm long)

Tiger beetle, *Cicindela semicincta*, **Carabidae** (14 mm long)

genus, *Cicindela*, are active in daylight in open sandy areas.

Predacious Diving Beetles (Family Dytiscidae)

This is one of two large aquatic families with 199 species in 42 genera. The hind coxae extend to the elytra on the sides and the antennae are filiform. They range from 1 to 40 mm in length, and are very smooth and mostly oval for rapid underwater movement. The hind legs are enlarged and the tarsi have a border of hairs acting as oars when the legs beat in unison. Dytiscids breathe by coming to the surface backwards and exposing the tip of the abdomen which draws in air to store under the elytra. Both adults and larvae feed on dead or living animal material ranging from insects to fish.

Different genera live in different positions in water bodies which separate them by size and shape.

They are found in most fresh water bodies, and most species can fly in search of another when these dry up. Brown and black are common colours of dytiscids, though some are more

Diving beetle, *Hydaticus parallelus*, **Dytiscidae** (20 mm long)

colourful like the 20-mm long *Hydaticus parallelus*.

Whirligig Beetles (Family Gyrinidae)
With only 20 species, this is a small but very distinctive aquatic family. Their flattened, streamlined bodies (4–18 mm in body length) have the middle and hind legs modified to form short oars, while the front legs are long and point forward for grasping. Their two eyes are effectively divided into four: two looking into the water and two above. Whirligigs spend most time on the surface swimming in groups, tracing quick circular patterns while looking for dead or injured insects to feed on. When threatened, they usually dive; but they can fly and some will come to lights, like the eastern *Macrogyrus elongatus*.

Suborder Polyphaga
Water Beetles (Hydrophilidae)
This family with about 198 species in 50 genera, ranging from 2 to 40 mm in body length, can be distinguished from dytiscids by their short antennae with a distinctive three-segmented, pubescent club, preceded by a smooth, widened segment (cupule). In aquatic

BEETLES

Whirligig beetle, *Macrogyrus elongatus*, **Gyrinidae** (12 mm long)

species, the cupule is used to transfer air to an area on the abdomen (plastron) which stores an air bubble. Their maxillary palps are longer than the antennae, as shown in Figure 38. Unlike the dytiscids, the adults are phytophagous or saprophagous, though most larvae are predators. It is a varied family which includes terrestrial leaf litter species and some that live in moist dung.

Family Histeridae
A small but distinctive family with 185 species in 41 genera. Histerids are characterised by their shiny, usually smooth and very hard body, small size

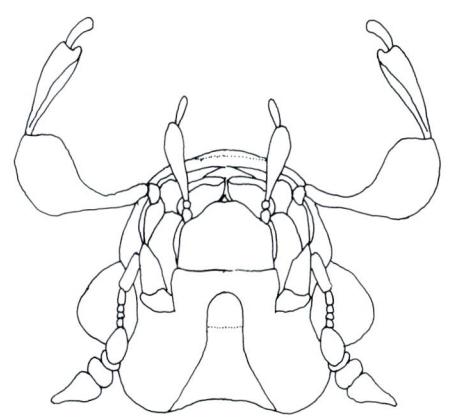

Figure 38. The underside of the head of *Hydrophilus pedipalpus* (family Hydrophilidae) showing the extra large maxillary palps pointing up and the intricate and short antennae pointing down. [R. Storey]

(1–12 mm in body length) and often flattened, occasionally cylindrical shape. Their mandibles are usually sharp and protruding. The antennae are clubbed and elbowed. The elytra are often shortened, with one or two abdominal segments visible from

BEETLES

above, and the legs can fit into depressions in the body.

Most species are active predators as adults and larvae. Some live in leaf litter, others in rotting wood or carrion and 40 species live in ant nests and possess special organs which attract the ants. The genus *Hololepta* includes the largest Australian histerids (Figure 39) which are very flat and found under loose bark of dead trees.

Rove beetle, *Pinophilus* sp., **Staphylinidae** (15 mm long)

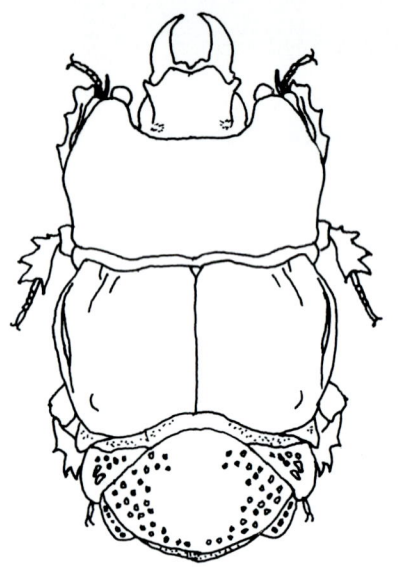

Rove beetle, *Actinus macleayi*, **Staphylinidae** (18 mm long)

Figure 39. *Hololepta* sp. (family Histeridae), 9 mm. [R. Storey]

Rove Beetles (Family Staphylinidae)

This is one of the major beetle families. Even though over 922 species in 231 genera are described, this is probably only half the Australian fauna. They range from 1 to 22 mm in body length, and the majority are recognised by having greatly reduced elytra with most of the abdominal segments not only visible from above but sclerotised and movable. Most are black or brown like the large genus *Pinophilus*.

The family is poorly known as most species are small, often cryptic and live as predators hidden in soil and leaf litter. Others are associated with dung, carrion or fungi. Among the predators are the largest and most colourful species, like the 20-mm *Actinus macleayi*. There are also many bizarre-shaped inquilines in Australia which inhabit the nests of ants and termites.

Family Pselaphidae

Pselaphidae (see Figure 40) is another poorly known family due to its members' very small size (1–4 mm in body length) and habit of living almost entirely in moist leaf litter. The 566

BEETLES

with ants and in animal nests and caves. Despite their small size, they often have marvellous pits, horns and hair patterns which reward study with a microscope.

Stag Beetles (Family Lucanidae)

Although only 94 Australian species in 17 genera are currently named, this is a well-known family. Lucanids (6–70 mm in body length) have forelegs adapted for burrowing in wood; and the last three segments of the antennae are lamellate. Brown is a common colour, like the genus *Cacostomus*, but there are many species which are bright metallic. The mandibles are sexually dimorphic, often enlarged in males. These antler-like mandibles give the family its common name, stag beetles. Lucanids usually develop in decaying fungus-infested logs, or in the humus beneath. Most are found only in the wetter areas of the east coast of Australia. Adults may feed on nectar, young foliage, sap flows of trees and ripe fruit. Despite being common in some habitats, their nocturnal habits make them less visible than might be expected for such striking insects.

Family Passalidae

The 34 species in nine genera of this distinctive family are largely confined to the wetter subtropical and tropical forests. They are large (20–60 mm in body length), slightly flattened, with a forward pointing head, often with a small horn. They have lamellate antennae, and a narrow waist between the prothorax and the elytra (which are strongly striated). Passalids are subsocial insects, their families living in decaying wood. Adults prepare the wood into a soft pulp which feeds the larvae and there is communication within the family group using

Figure 40. *Malleecola myrmecophila*, (family Pselaphidae), 2 mm. [R. Storey]

described species in 163 genera are probably less than half the actual Australian fauna. The elytra are shortened, exposing a non-flexible abdomen (unlike the rove beetles). They are usually reddish-brown with reduced two- to three-segmented tarsi, and antennae with a pronounced three-segmented club. From what little is known, they mainly prey on small arthropods.

Other than leaf litter dwellers, there are many species which live

Stag beetle, *Cacostomus squamosus*, **Lucanidae** (22 mm long)

115

BEETLES

Mastachilus sp., **Passalidae** (55 mm long)

stridulation. The larger species are found in the genus *Mastachilus*.

Family Scarabaeidae

This is the second largest family of beetles with 2216 described species in 270 genera, which is probably only about two-thirds of the actual Australian fauna. Scarabaeids are small to large (2–70 mm in body length), with mandibles often hidden from above by a plate called the clypeus. At least the last three segments of the antennae form a lamellate club. The body is usually stocky, and sometimes brightly coloured. Males in some genera have prominent horns on the head and/or pronotum. The 'C'-shaped larvae are white with a prominent dark head and three pairs of legs. Most larvae feed on fresh or decaying vegetable matter, often in the soil. Consequently the adults have the front legs adapted for burrowing.

The family is divided into seven distinct subfamilies in Australia. Scarabaeinae species and many Aphodiinae species are mostly dung feeders as adults and larvae. Some scarabaeines roll balls of dung for

Dung beetle, *Cephalodesmius armiger*, **Scarabaeidae** (13 mm long)

OPPOSITE: Chafer, *Polystigma punctata*, **Scarabaeidae** (18 mm long)

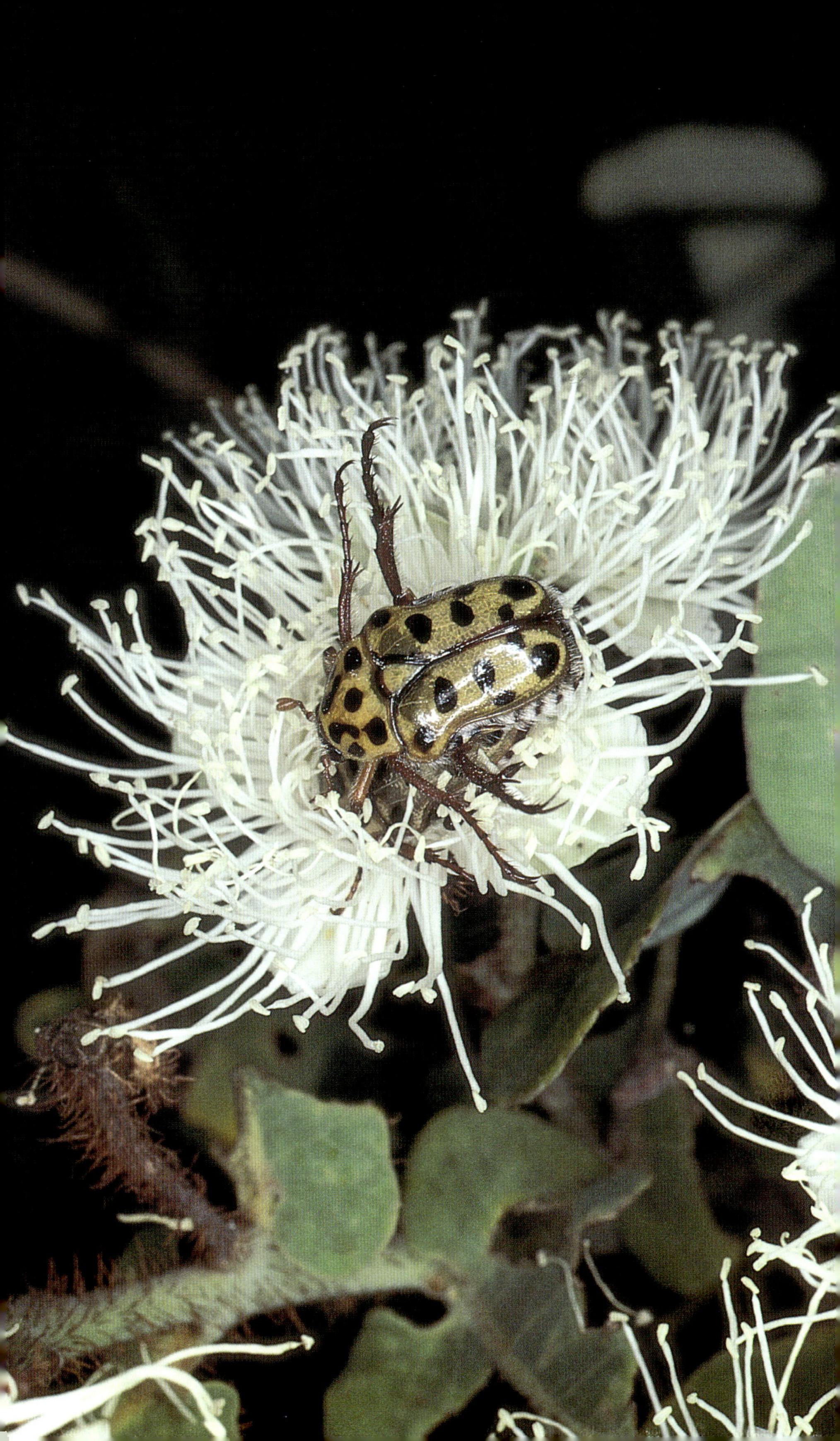

BEETLES

ABOVE: Green Christmas beetle, *Calloodes rayneri*, **Scarabaeidae** (24 mm long)

LEFT: Furrowed rose chafer, *Trichaulax macleayi*, **Scarabaeidae** (28 mm long)

BELOW: Nectar scarab, *Phyllotocus apicalis*, **Scarabaeidae** (8 mm long)

adult food and to provision brood nests for their larvae. Their mouthparts are soft as an adaptation to a liquid diet.

The Cetoniinae are flower scarabs, feeding on nectar as adults and are often brightly coloured. They are strong fliers and have the elytra modified so the beetle can fly without raising them. This leads to a loud buzzing noise in flight. A common species on blossom is the 15-mm long *Polystigma punctata*, while in the tropics lives the very distinctive *Trichaulax macleayi*.

The subfamily Rutelinae includes the colourful Christmas beetles which feed on grass roots as larvae and can defoliate trees as adults. The large genus *Anoplognathus* has mainly brown species and *Calloodes* mainly green ones. Dynastinae species are usually black in colour and many males have prominent horns. The largest Australian scarabaeid, the 'rhino' beetle of the wet tropics, belongs in this group.

The largest subfamily is Melolonthinae, commonly called chafers. They are usually brown to black in colour, often feed on grass roots as larvae and have mostly nocturnal adults. Among the diurnal species are the small, blossom-frequenting *Phyllotocus*. Many, such as the cane beetles, are agricultural pests.

Click Beetles (Family Elateridae)

This is a large family with about 800 species in 67 genera. Elaterids are well known due to the unique clicking mechanism of the adults. On the underside of the prothorax there is a spine which fits into a pit in the mesothorax. When upside down, the beetle bends its body to half insert this spine so it rests on the edge of the cavity. The sudden snap as it enters the cavity straightens the body so violently that it somersaults high

Predatory click beetle, *Paracalais* sp., **Elateridae** (25 mm long)

into the air, helping to confuse a predator. Other features are the elongate, somewhat flattened body (4–50 mm in body length), with antennae that are slightly to strongly serrate and a visible labrum. Many species can be seen around lights at night, others can be seen on flowers. The larvae are long with a hardened shiny, very cylindrical body which gives them the name wireworms. Most are plant feeders, including roots, which makes some agricultural pests. Others are predacious, like *Paracalais*, which hunt other wood-boring beetles.

Family Lycidae

A medium-sized family of distinctive soft-bodied beetles. The 200 or so species are placed in 14 genera. Features are the soft, flattened body (5–20 mm in body length), and elytra and pronotum with patterns of ridges. They are usually coloured brick-red and black, and the antenna) segments are serrate, flattened and wide. Adults frequent foliage and blossom. Their colour advertises that they are distasteful to predators. Because of this they are the model for many examples

BEETLES

Lycidae, probably *Metriorrhynchus* sp. (6 mm long)

of mimicry in other families of beetles and even wasps, flies and moths (see page 20). The largest genus is *Metriorrhynchus*.

Soldier Beetles (Family Cantharidae)

A small but distinctive family of flattened, soft-bodied beetles (3–15 mm in body length), similar to the Lycidae, but with filiform instead of serrate antennae. Most have gold and black colour combinations; though the commonest species, which sometimes covers entire blossoming trees, is the dark metallic *Chauliognathus lugubris*.

Fireflies (Family Lampyridae)

A small family (26 species in four genera) of well-known beetles (4–12 mm in body length). Their most apparent characteristic is the strong flashing luminescence in the adults, with the light-producing organs located near the tip of the abdomen. In females, it is only on the fifth segment, but in males it may be on the fifth and sixth. Other characters are the soft, flattened body, very large eyes which are partly concealed from above, and short antennae. The larvae are only slightly luminous and feed on snails. The males often flash to the females as a mating ritual and

Plague soldier beetle, *Chauliognathus lugubris*, **Cantharidae** (18 mm long)

BEETLES

Firefly, *Luciola* sp., **Lampyridae** (9 mm long)

Large auger beetle, *Bostrychopsis jesuita*, **Bostrichidae** (22 mm long)

Clerid, *Phlogistus* sp., **Cleridae** (10 mm long)

synchronised flashing by groups of males is known. Most species are found in the wet forests and mangroves of the east coast of Australia and *Luciola* is the major genus.

Auger Beetles, Powderpost Beetles (Family Bostrichidae)

The family of auger beetles, recently merged with the Lyctidae, includes about 55 species in 24 genera. Larvae and adults are borers in dead and dying wood and include species that are pests of timber and furniture. Most species are small (usually less than 10 mm long, ranging from 2 to 22 mm) and elongate and cylindrical. The adults often have spines on the head, the pronotum and the rear of the ely-tra. Antennae have a loose club and the eyes are very round and protruding. The typical form has a very domed pronotum, with the head mounted below and pointing down, as in the largest species, *Bostrychopsis jesuita*, which lives on eucalypts and acacias.

Family Cleridae

A medium-sized family (350 species in 57 genera) of distinctive beetles, most of which are active predators both as larvae and adults (5–40 mm in body length). The adults are usually subcylindrical and elongate, with strong running legs and large eyes. They are

121

BEETLES

usually hairy, with patches of erect and recumbent setae. Clerid genera tend to have an appearance characteristic for the type of habitat they inhabit. Some brightly coloured or metallic genera, like *Phlogistus* are common on flowers. More cryptically coloured genera are often found on logs or tree trunks. Others inhabit leaf litter, or are associated with fungi or even termite mounds.

Jewel Beetles (Family Buprestidae)

This is a major and very distinctive family in Australia with over 1200 species in 78 genera, including several hundred species in the largest genus *Castiarina*. Adults are elongate and usually flattened (3–65 mm in body length). They have large eyes, short antennae and legs which can be retracted against the body. They are often very brightly coloured and sometimes metallic. The larvae are usually wood borers in branches, stems or roots, though some are leaf miners or gall formers. Adults, particularly the brightly coloured ones (the large

Jewel beetle, *Castiarina carinata*, **Buprestidae** (14 mm long)

genera *Curis*, *Stigmodera* and *Castiarina*), are nectar feeders, especially on eucalypt blossom. Others, especially in the subfamily Chalcophorinae which includes the large green *Cyphogastra*, are leaf feeders. This family is very popular with amateur collectors.

Jewel beetle, *Cyphogastra farinosa*, **Buprestidae** (30 mm long)

BEETLES

Adult and pupa of ladybird, *Illeis galbula*, Coccinellidae (7 mm long)

Ladybirds (Family Coccinellidae)

This family, with about 300 species in 54 genera in Australia, is one of the most economically important groups of beetles. Ladybirds (1–15 mm in body length) are circular, convex beetles, with the antennae and legs short and usually hidden beneath the body. Although most species are very small, black or brown with paler markings and covered with short recumbent hair, the more familiar species in gardens are smooth and shiny with black and orange or yellow colouring, like *Illeis galbula*. All but the genus *Epilachna* are predacious as adults and larvae, often on important agricultural pests like mites, aphids and scale insects which are difficult to control with chemicals. The Australian genera *Rodolia*, *Cryptolaemus* and *Rhyzobius* have become important biocontrol agents of Australian scale insects that have become pests overseas.

Darkling Beetles (Family Tenebrionidae)

This large family is one of the most diverse in Coleoptera with over 1500 Australian species in 203 genera and eight subfamilies. This diversity makes it hard to characterise the family features which include body length ranging from 2 to 30 mm; simple moniliform antennae; usually dull brown to black in colour, in some cases metallic; and the fourth and fifth abdominal sclerites have a membrane at the base. Adults and larvae usually feed on dead plant or fungal material, though some are agricultural pests. The genera *Palorus*, *Tribolium* and *Tenebrio* (which includes the mealworm beetles) are pests of stored grains. The family is found in most habitats from deserts to rainforests. They can be found on decaying wood, in the nests of vertebrates and on beaches and sand dunes. The desert (including the pie-dish beetles) and stored-grain species are adapted to very dry conditions. *Gonocephalum* is a widespread genus with several crop pest species.

Longicorn Beetles (Family Cerambycidae)

This large family is currently estimated at 1200 species in five subfamilies and 314 genera. They are mostly elongate and flattened to rounded (3–80 mm in body length). The antennae are long, from three-quarters to several body lengths

False wireworm beetles, *Gonocephalum* sp., Tenebrionidae (8 mm long)

ABOVE: Longicorn beetle, *Xixuthrus microcerus*, **Cerambycidae** (75 mm long)

LEFT: Longicorn beetle, *Aphiorhynchus* sp., **Cerambycidae** (16 mm long)

ABOVE: Longicorn beetle, *Penthea pardalis*, **Cerambycidae** (30 mm long)

Left: Leaf beetle, *Polysastra costatipennis*, **Chrysomelidae** (12 mm long)

Leaf beetle, *Paropsis obsoleta*., **Chrysomelidae** (10 mm long)

in size and they appear to be mounted within the margins of the large eyes. The mandibles are obvious and strong, and the legs are long with distinct claws. The larvae are usually white grubs boring in dead or dying wood. Some are agricultural pests, including species that ringbark tree branches to provide a medium for their larvae.

The subfamily Prioninae are medium to large flattened beetles with large protruding mandibles, like the 75-mm *Xixuthrus microcerus* from north Qld. The larvae are a favourite Aboriginal food.

The subfamily Cerambycinae are flattened, smaller and more elongate with gangly legs. They are very diverse with most colours and even metallic hues, as found among the species of *Aphiorhynchus*.

The more convex and stout species of the subfamily Lamiinae have deflexed heads and include the largest Australian beetle, the 80-mm *Batocera wallacei*, found in Cape York Peninsula. The distinctive

Penthea pardalis is 40 mm in length and lives on wattles.

Leaf Beetles (Family Chrysomelidae)

This is one of the largest families of beetles with approximately 3000 Australian species in 273 genera. The boundaries of this family vary greatly between authors. Lawrence and Britton (1991) recognise 12 subfamilies including the Bruchinae or seed-weevils. Leaf beetles are mainly small (usually 5–15 mm in length, but ranging from 1 to 30 mm); and present in many shapes: flattened, globular, or elongate. They are usually smooth and often colourful; the tarsi are lobed and large for the body size; and the antennae are generally one-third to one body length. Nearly all adults and larvae feed on living plant tissue, so the family contains many economically important species, both pests and beneficials in biocontrol.

The subfamily Sagrinae contains the largest species, in the genus *Sagra*, with very large hind legs. Swollen hind femurs are also a feature of the subfamily Bruchinae, whose larvae are seed feeders, and some Galerucinae which are known as flea beetles for their jumping abilities. A common non-jumping galerucine is *Polysastra*. The Hispinae include some strikingly spiny beetles in the genus *Hispellinus*. The Chrysomelinae is the largest subfamily and includes the very many species of the genus *Paropsis* and its relatives, with characteristic domed shapes.

Weevils (Family Curculionidae)

This is the largest beetle family in Australia with an estimated 8000 species in over 692 genera spread among numerous subfamilies which are currently being redefined. Despite great diversity, most share the distinctive characteristic of a rostrum, an elongate snout which extends the

BEETLES

mouthparts. They range in size from 1 to 60 mm. The antennae have a long scape (which gives them a distinct elbow) and often a distinct club; the legs can be retracted against their often very hard bodies. The larvae are 'C'-shaped, legless and usually feed on wood and other plant parts. Many weevils are economically important pests of trees, vegetables, grains and ornamentals, attacking the stems, roots, seeds and fruits and sometimes the leaves. There are few crops that are not affected by one or more weevil; although, there are some beneficial species, including some used in bio-control of weeds.

Some interesting genera include the stout, ground dwelling *Phalidura* with large abdominal pincers on the males; *Bagous* and others which feed on aquatic plants; and *Rhynchaenus* which are small jumping weevils. Probably the most typical weevils are in the large genus *Leptopius* which often feed on wattles.

There are two subfamilies which have tiny beetles that have no obvious

Pinhole borer weevil or ambrosia beetle, **Curculionidae–Platypodinae** (8mm long)

Weevil, *Leptopius quadridens*, **Curculionidae** (20 mm long)

Rhinotia hemistictus, **Belidae** (15 mm long)

rostrum. The Scolytinae are known as bark beetles and are known for their characteristic complex branching tunnels under bark. The Platypodinae make tiny tunnels straight into the heartwood and are known as pinhole borers, or as ambrosia beetles for the fungi which they cultivate in the tunnels for food.

Family Belidae

This is a small weevil family but is very well represented in Australia with around 175 species. They have a very characteristic elongate, cylindrical form (5–25 mm in body length).

They also have a rostrum and sometimes are patterned with spots and stripes. The main difference between these and true weevils is their antennae which do not have an elbow. Adults are usually found on plants, with the larvae boring into the branches. *Rhinotia hemistictus* is typical, it is 20 mm long and feeds on wattles.

Family Brentidae

With 200 species in 30 genera, this family also includes the former family Apionidae which makes the family definition less distinct. The Apioninae are small (less than 10 mm in length),

Ectochemus decimmaculatus, **Brentidae** (25 mm long)

often black, seed feeders and have a large, very round abdomen with often very long and narrow rostrum. The larger wood-feeding Brentinae (2–40 mm in body length) are usually very narrow and elongate, and are more often brown than black. Both groups usually have clubbed antennae but with no elbow which distinguishes them from the true weevils. Brentinae are often very sexually dimorphic with the males having elaborate and sometimes flattened rostrums and other differences. The tropical genus *Ectochemus* has some of the few colourful species.

Further Reading

Crowson, R.A. (1981), *The biology of the Coleoptera*, Academic Press, London.

Hawkeswood, T. (1987), *Beetles of Australia*, Angus and Robertson.

Lawrence, J.F. and Britton, E.B. (1991), 'Coleoptera' (ch. 35) in *The Insects of Australia*, CSIRO, Melbourne.

Lawrence, J.F. and Britton, E.B. (1994), *Australian Beetles*, CSIRO, Melbourne University Press, Melbourne.

Lawrence, J.F., Hastings, A., Dallwitz, M.J. and Paine, T.A. (1999), *Beetle Larvae of the World*. Interactive identification and retrieval for families and subfamilies. CD-ROM and manual published by CSIRO Publishing.

Lawrence, J.F., Hastings, A., Dallwitz, M.J., Paine, T.A. & Zurcher E.J. (1999), *Beetles of the world. A Key and Information System for Famiies and Subfamilies*. CD-ROM, CSIRO Publishing.

Matthews, E.G. (1980–2002), *A Guide to the Genera of Beetles of South Australia*, Special Education Bulletin Series, South Australian Museum, Adelaide.

Moore, B.P. (1980–92), *A Guide to the Beetles of South-Eastern Australia*, Australian Entomological Press.

STYLOPIDS
ORDER STREPSIPTERA

World Families: 8; species: 530 **Australian families:** 6; species: 160

Characteristics of Order

Complete life cycle; endoparasitic insects with extreme sexual dimorphism. The males are mobile with forewings reduced to elongate knobs and the hind wings are membranous, very broad and with little venation; the mouthparts are reduced and nonfunctional; the antennae are often multibranched (see Figure 41). The females are larviform with no wings and usually no legs; the mouthparts are greatly reduced; and they do not leave their hosts except in one family.

Males superficially resemble the Rhipiphoridae beetles and have been considered a family (Stylopidae) of the beetle order in the past. An obvious difference is the club-like forewings of stylopids.

Life History

Strepsiptera are endoparasitic on other insects, especially true bugs (order Hemiptera) and wasps (Hymenoptera). This lifestyle explains their very modified bodies. The females do not leave the host (except the genus *Eoxenos*), producing large numbers of minute first instar larvae. These larvae have long, thin legs and are mobile, some

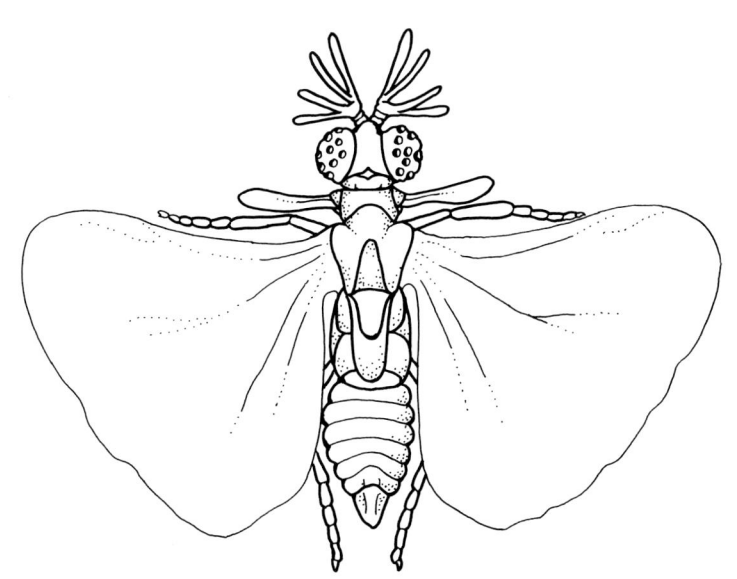

Figure 41. A typical male stylopid (family Mengenillidae) with a wingspan of about 6 mm. [R. Redman]

jumping onto new hosts. On entering the body of the host, the larva moults into a legless second instar which moults through several more instars without leaving the host. In one family, the Mengenillidae, both sexes leave the host to pupate externally; but in most, the larvae half emerge and pupate on the host without killing it. The winged males emerge and, probably through specific pheromones, seek the females which emerge and stay on the living host. After mating, the females produce more live first-instar larvae.

Biology

Strepsipterans are seldom seen because of their small size (male wingspan: 1–8 mm) and parasitic habits, but males sometimes come to lights. All species parasitise a variety of orders and there is even a family (Myrmecolacidae) where the males and females parasitise different insect orders. Usually only one individual lives on a host, though two individuals and even two different species have been recorded. Most species are not confined to a single host species, but rather a selection within a specific family or order. While most hosts do not die, they are usually rendered sterile. As some host species are pests, some strepsipterans are considered beneficial.

Classification

There are two suborders. The suborder Mengenillidia includes the one family, Mengenillidae, with free-living females, which has 20 Australian species. The other five families are in the suborder Stylopida, where Stylopidae is the largest family with 65 species. Identification is difficult and based mainly on male characteristics because, in many species, the females have not yet been recorded.

Further Reading

Kathirithamby, J. (1989), 'Review of the Order Strepsiptera', *Systematic Entomology*, vol. 14, pp. 41–92.

Kathirithamby, J. (1991), 'Strepsiptera' (ch. 36) in *The Insects of Australia*, CSIRO, Melbourne.

SCORPION FLIES AND HANGING FLIES
ORDER MECOPTERA

World families: 9; species: 500 Australian families: 5; species: 27

Characteristics of Order
Complete life cycle; small to medium (wingspans up to 50 mm); head has a beak-like projection or rostrum, with mandibulate mouthparts at the tip; two almost equal, membranous, strongly veined pairs of wings (one species is wingless); long legs with long, slender tarsi and large claws in some; males sometimes have a highly modified and upturned last abdominal segment reminiscent of a scorpion sting. Larvae are caterpillar-like with short legs and no prolegs, and have well-developed mandibles.

Life Cycle
Life histories are varied. In some predacious species, the males offer a nuptial meal to attract females to mate. The eggs are dropped to the ground or more often placed in chambers in moist soil. The larvae are free living in moist environments such as leaf litter, moss and near water, where they are usually scavengers on dead insects. The first three instars live about a month and are followed by an inactive prepupal stage within the soil that can last several months. Pupation then takes place in soil chambers.

Biology
Many adult mecopterans are carnivorous, with some herbivores and scavengers. The species of Bittacidae, the largest family, are active predators. They hang from twigs by their forelegs and use the long clawed tarsi of their hind legs to capture passing prey. Less is known of the habits of other families. Most species are restricted to wetter areas of eastern Australia from Tasmania to northern Qld, with only four species in southwest WA.

Classification
This is a small order in which the Australian species are mostly endemic and one family, the Choristidae, is restricted to Australia.

Family Bittacidae
This is the largest family with 14 species. They are characterised by very long hind tarsi which end with an enlarged claw for catching prey. Many species are rare, with some of the 10 species of *Harpobittacus* being the most common. They are the largest mecopterans and tend to be reddish in colour. Their general appearance is similar to that of crane flies but the 'beak' mouthparts at a glance, and the tarsal claw and the presence of two pairs of wings in detail, easily distinguish the two groups (true flies have only one pair of wings). Adults are slow flying and usually hang in vegetation ready to strike.

Family Apteropanorpidae
The most unusual mecopteran is *Apteropanorpa tasmanica*, the single species of the family Apteropanorpidae. It is about 10 mm long, wingless, with

Scorpion fly, *Harpobittacus tillyardi*, **Bittacidae** (30 mm long)

long antennae and upturned bulbous genitalia in the males. They scamper about and feed on moss in Tasmania and are active even in the snow. Similar species in Europe are known as snow fleas.

Family Choristidae
This family of eight species can be recognised by not having the enlarged tarsal claws of bittacids. They tend to live near water and are apparently mainly herbivores.

Family Meropeidae
In WA, there are two species in the Meropeidae which differ from the bittacids found there by having brown-coloured wings that overlap at rest.

Further Reading
Byers, G.W. (1991), 'Mecoptera' (ch. 37) in *The Insects of Australia*, CSIRO, Melbourne.
Byers, G.W. and Thornhill, R. (1983), 'Biology of the Mecoptera', *Annual Review of Entomology*, vol. 28, pp. 203–28.

FLEAS
ORDER SIPHONAPTERA

World families: 16; species: 2380 Australian families: 9; species: 90

Characteristics of Order
Complete life cycle; minute to small (1–10 mm in length); wingless; strongly laterally compressed; piercing and sucking mouthparts for a life as ectoparasites on mammals and birds; hind legs are highly modified for jumping and with various setae and spines for holding onto the hair or feathers of their hosts. Larvae are wormlike, and live in the nests of their hosts.

Life Cycle
Most fleas lay several hundred eggs per female, either in the nest of the host or on the host from where they usually fall into the nest. The larvae, with three instars, live in the nest from several weeks to many months, feeding on organic matter. Only one species has a parasitic larvae which burrows into the skin of the host. The larvae pupate within various shaped cocoons in the nest and adult emergence can be delayed up to several months until a suitable host is present. The vibration of a host arriving is enough to trigger the emergence of adult fleas.

Biology
All adult fleas intermittently feed on the blood of mammals or birds. They either live on the host or spend more time in the nest. About 95% of fleas feed only on mammals and most will feed on more than one host species. Some fleas can feed on dozens of host species and a mammal may have several species of fleas attacking it at any one time. Fleas occur in nearly all Australian habitats, including the subantarctic islands, and most are endemic. Many species are of economic importance. Most fleas which are associated with humans and domesticated animals like cats and dogs have been accidentally introduced. However, the rabbit flea was introduced deliberately to aid the spread of myxomatosis. Several human diseases are spread by fleas, including bubonic plague and murine typhus.

The fleas most likely to change hosts to deliver these diseases to humans are the rat fleas (*Xenopsylla cheopis*) which are found all over the world. The cat flea and the dog flea also bite people, but these species generally are only irritating rather than dangerous. The human flea (*Pulex irritans*) even implies this in its Latin name as flea bites often lead to allergic reactions. More serious is the role of dog fleas in the life cycle of tapeworms that may infect children.

Classification
Dunnet and Mardon (1974) give a good introduction and a key to the 33 genera of fleas present in Australia. Body-plate arrangements and patterns of

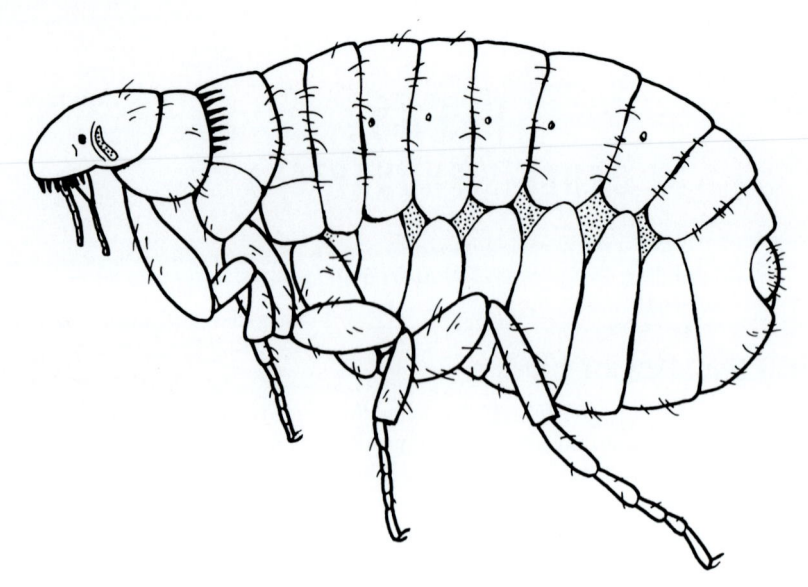

Figure 42. The dog flea, *Ctenocephalides canis* (family Pulicidae), which occurs worldwide. [R. Redman]

setae are used to separate the families, all of which are microscopic characteristics.

The largest family is the Pygiopsyllidae which includes about half the endemic species which have marsupial, rodent and bird hosts. The family Pulicidae contains the introduced and worldwide human flea (*Pulex irritans*), dog flea (*Ctenocephalides canis*) (Figure 42) and cat flea (*C. felis*), as well as many endemic species.

Further Reading

Dunnet, G.M. and Mardon, D.K. (1974), 'A monograph of Australian fleas (Siphonaptera)', *Australian Journal of Zoology*, suppl. 30, pp. 1–273.

Dunnet, G.M. and Mardon, D.K. (1991), 'Siphonaptera' (ch. 35) in *The Insects of Australia*, CSIRO, Melbourne.

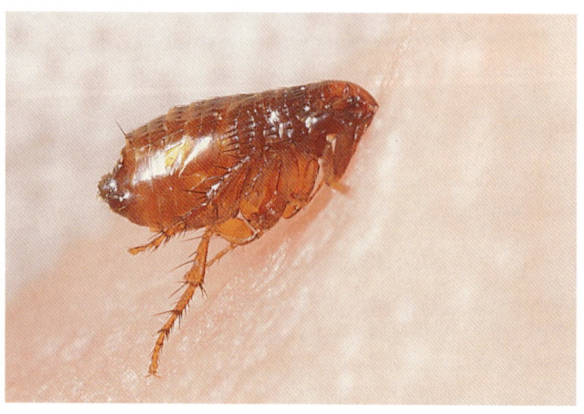

Cat flea, *Ctenocephalides felis*, **Pulicidea** (2 mm long)

FLIES
ORDER DIPTERA

World families: 130; species: 150 000 Australian families: 98; species: 7786

Characteristics of Order

Complete life cycle; minute to large insects with membranous forewings and the hind wings reduced to small, club-shaped halteres; head has modified sucking and sometimes piercing mouthparts, large compound eyes and usually very short simple antennae (branched and frilled in mosquitoes and crane flies); legs are often long with five-segmented tarsi. Larvae are the typical maggots with no true legs and very reduced heads.

Possible confusion may arise between flies and the few other insects which have one pair of wings (e.g. some male scale insects (Hemiptera) and some mayflies (Ephemeroptera). However, flies are the only insects that have hindwing halteres. Some flies are also good mimics of wasps. These can also be easily distinguished in the hand because flies have only two wings and not the four of wasps.

Life Cycle

Being strong fliers many flies mate on the wing. The eggs are usually laid singly on or close to the larval food source. They hatch quickly, in some families hatching within the female so that live larvae are deposited. In extreme cases (like the African tse-tse fly), the larvae live inside the mother being fed by special glands, and only emerge to pupate. Fly larvae are termed maggots, and are simple, pale and legless (see Figure 43). The more advanced fly larvae, like house flies, have minute reduced heads, while the primitive forms, like mosquitoes and crane flies, have larger heads. With so many families, a large variety of food media are utilised. Usually this is moist, such as decaying animal and plant matter. There are also various forms of parasitism and water filter-feeding.

There are usually four larval instars. Some species can grow to an adult within a week. The larvae pupate within the host or other food supply, or in the ground nearby, sometimes forming a puparium or spinning a cocoon. Some flies have a special sac-like structure (ptilinum) on the head which is used to break out of the puparium. Many aquatic larvae pupate in water as an active swimming stage.

Biology

In such a large order many different habitats are exploited. The larvae of most species feed on moist, often decomposing food items such as carrion, fungi, dung and rotting vegetable matter. Many species are parasitic on the larvae of other insect orders. Others live in fresh or brackish waters, usually as filter feeders of fine organic matter.

FLIES

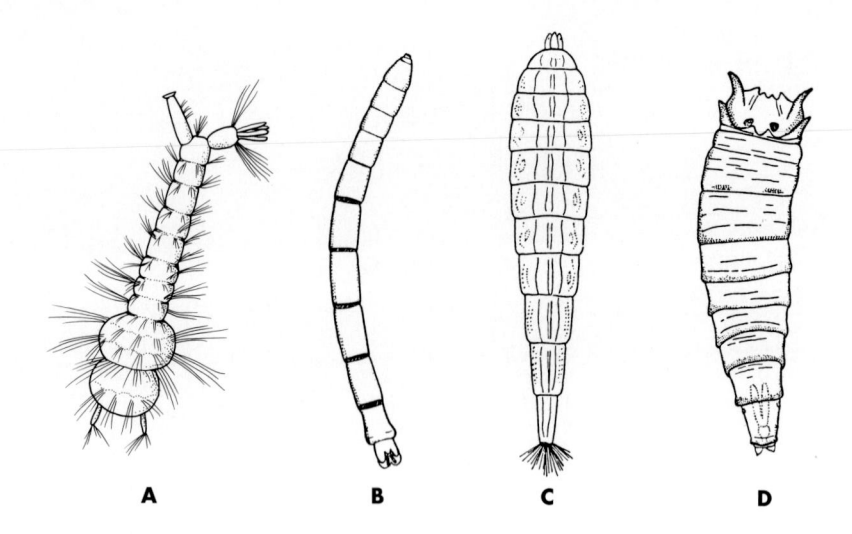

A   B   C   D

Figure 43. Some fly larval types: (a) the mosquito 'wriggler' (family Culicidae) which is active underwater; (b) an elongate crane fly larva (family Tipulidae) which lives in swampy conditions; (c) a larva of a soldier fly (family Stratiomyidae) which lives in damp earth; (d) a 'maggot' larva of a blowfly (family Calliphoridae) living in rotting meat. [after A. Tonnoir]

Adults are usually strong fliers and diurnal, though there are a number of well-known nocturnal species including some mosquitoes. The adult mouthparts are designed for lapping up moist food, or for piercing and sucking. In most species digestion is partly external: enzymes are introduced into the food, and then the softened product is mopped up. Blood-sucking groups like mosquitoes and march flies first pierce the skin of the prey either with a sharp pointed or rasp-like tip to their proboscis.

All environments are exploited by flies, with forests and the margins of water bodies having the greatest species diversity. Even in the arid interior, the bush flies (Muscidae), can build up to enormous numbers during favourable seasons.

Classification

As in many groups of insects, the classification of Diptera is in a state of change, and most opinions on family relationships will be disputed by other dipterists. Colless and McAlpine (1991) recognise two suborders: the more primitive Nematocera has 19 families in six divisions and about one-third of all the species; the remaining 79 families are placed in Brachycera with two divisions.

Members of the suborder Nematocera have the antennae usually filiform, with eight or more segments which are sometimes plumose (featherlike) in males, and are usually longer than the thorax. Wing venation has the CuA vein rarely converging with the 1A vein, leaving an open cell to the wing margin (see Figure 44). Adults are usually slender with long legs; and larvae are usually aquatic or subaquatic with some gall makers (e.g. mosquitoes and crane flies).

Members of the suborder Brachycera have the antennae short to very short with less than seven segments usually consisting of a broad base and a narrow bristle called an arista. In wing venation, the CuA vein

FLIES

converges towards the 1A vein, often meeting and forming a closed cell that does not reach the wing margin (see Figure 44). Adults are heavier set with relatively short legs (e.g. house flies and march flies).

Suborder Nematocera

Crane Flies (Family Tipulidae)

These are small to large flies (wingspan: 6–75 mm) with slender bodies and very long legs. The thorax has a 'V'-shaped pattern across it; the ocelli are absent; and the wings do not have scales along the veins and margins. This is the largest family of flies in Australia with 704 species. The greatest diversity is in south-eastern Australia, often along creeks and one of the largest genera is *Gynoplistia*. The showy genus *Nephrotoma* has a worldwide distribution. The larvae are mainly aquatic with some found in damp soil and others in rotting vegetation. Most adults drink but do not feed.

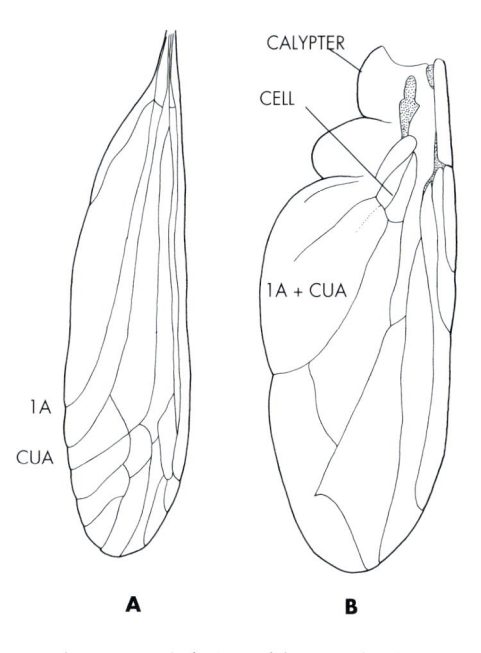

Figure 44. Typical wings of the two suborders of flies: (a) Nematocera (family Tipulidae) showing the separate veins 1A and CuA, reaching the margin of the wing; (b) Brachycera (family Muscidae) showing the veins 1A and CuA joined, creating an extra cell not found in Nematocera.

Crane fly, *Gynoplistia* sp., **Tipulidae** (16 mm long)

Crane fly, *Nephrotoma australasiae*, **Tipulidae** (15 mm long)

Mosquitoes (Family Culicidae)

These are small flies with long narrow wings with scales along the veins and wing margins. The mouthparts are formed into a long proboscis and the antennae in males are usually plumose. The larvae are aquatic and known as wrigglers and are often seen coming to the surface to breathe through a syphon. They especially inhabit still water, often in small spaces like tree holes or even tin cans. The pupae are also aquatic and active. The adult females of most species feed on blood; the males never do so. The 275 Australian species feed at different times of the day and some are host specific. Generally they attack most vertebrates, even some fish. Feeding females can swell to twice their normal size like *Aedes*. Many mosquitoes are important disease carriers, acting as vectors for malaria (spread by *Anopheles* sp.), dengue fever (spread by *Aedes aegypti*), filariasis (spread by *Culex fatigans*) and other parasites and viruses. The majority of species are tropical, and many are also distributed outside Australia.

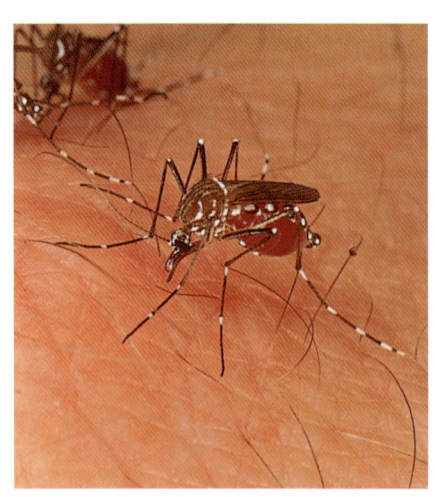

Dengue mosquito, *Aedes aegypti*, **Culicidae** (8 mm long)

FLIES

Midges (Family Chironomidae)

These are small (wingspan: 1–8 mm), nonbiting mosquito-like flies with 202 species. They differ from mosquitoes by having a humped thorax that almost overhangs the head and by not having a biting proboscis. The larvae are mostly aquatic, living on the bottom or on plants, and some are partly or wholly marine. The red-coloured bloodworms, found in muddy freshwater situations, are midge larvae which have a high haemoglobin content to aid living in oxygen-poor conditions. Their tunnel building can be a problem in irrigated areas. Adult midges often swarm around creeks and ponds and, like mosquitoes, are most abundant in the tropics.

Sand Flies (Family Ceratopogonidae)

These are the minute biting midges that often make seasides and many inland waterways very unpleasant. Most of the 174 species have wingspans less than 5 mm. They differ from the chironomids by having a biting proboscis, and by sitting with their wings flat over the body instead of upright. The tiny larvae are aquatic and many can live in brackish water.

Family Bibionidae

There are 32 species of these distinctive slow-moving, usually dark-coloured flies that are stouter than other nematocerans. They have short antennae inserted below the eyes, spines on the front tibiae and visible ocelli. In the males, the eyes are completely divided into an upper and lower pair. The larvae live in soil and decaying plant matter. The orange and black, apparently wasp-mimicking species of the genus *Plecia* can be common in the warmer north of Australia.

Suborder Brachycera

Horse Flies and March Flies (Family Tabanidae)

There are 243 species of these small to large (6–20 mm in body length), stockily built flies. They have large eyes, which often show iridescent reflections, that always meet in the middle in the males. Other characteristics

Plecia ornaticornis, **Bibionidae** (13 mm long)

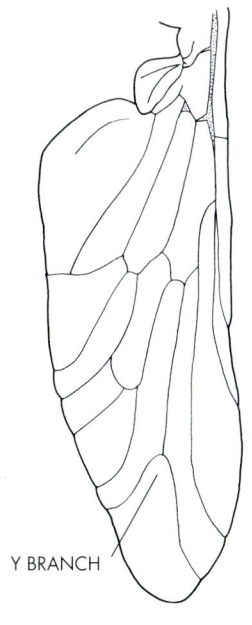

Figure 45. Wing of typical march fly (family Tabanidae), showing the 'Y'-shaped branching of veins at the wing tip. [R. Redman]

FLIES

March fly, *Dasybasis* sp., **Tabanidae** (25 mm long)

include three instead of the usual two pads at the tips of the tarsi; a long third antennal segment made up of fused parts; and the divergence of wing veins R4 and R5 to form a 'Y' across the wing tips (Figure 45). The males feed only on nectar; but the females, with a well-developed proboscis, are blood feeders. Many species bite humans, with the summer southern alpine species and many tropical wet season species being especially annoying. The larvae can be aquatic, living among floating vegetation, or in other damp and muddy habitats. *Dasybasis*, with 75 species, is the largest genus and all have hairy eyes.

Robber Flies (Family Asilidae)

A large distinctive family (640 species) of small to large hunting flies. The head often has a distinct 'neck' region and prominent eyes with a groove between them at the top of the head. They have three ocelli and a 'bearded' face of distinct setae around the stout and pointed proboscis. The thorax and legs are also very hairy, and the abdomen can be either short and stout or long and thin. Robber flies are active predators, catching other insects on the wing. Some genera such as *Blepharotes* are among the largest flies, with wingspans up to 75 mm. Among the more usual dark forms, there are colourful species with yellows and reds, some of which are wasp mimics. The genus *Laphria* has stout

March fly, *Cydistomyia* sp., **Tabanidae** (18 mm long)

Top: Robber fly,
Asilidae–Asilinae
(20 mm long)

Right: Robber fly, *Laphria* sp., **Asilidae** (25 mm long)

Below: Bee fly, *Bombylius* sp., **Bombyliidae** (25 mm across)

141

FLIES

Bee fly, *Exoprosopa* sp., **Bombyliidae** (20 mm long)

flies with extra hairy legs, while a more typical shape is the thinner more cylindrical body of the largest subfamily, the Asilinae. Adults tend to frequent forests, where they often perch on the tips of thin branches. Little is known of the larval habits. They live in soil and detritus and some are also predators.

Bee Flies (Family Bombyliidae)

There are 391 species of these small to large flies known for their distinctive habit of hovering between short bursts of horizontal flight. They have stout bodies, with the thorax and abdomen having no obvious join, and are hairy all over the body, with long thin hairless legs. The wings are long and held flat out from the body at rest. Adults hover near flowers on which they feed. Species are found all over Australia with many adapted to the arid interior, such as the species of *Exoprosopa* from SA. The larvae are parasitic on the eggs and larvae of several insect orders. Species of the large genus *Comptosia* have wingspans up to 70 mm.

Family Dolichopodidae

This is a large family of 320 species of small to medium-sized flies (wingspan: up to 17 mm). Their most distinctive

Longlegged fly, *Sciapus* sp., **Dolichopodidae** (6 mm long)

features are a metallic green to bronze body colour and long, slender legs. The antennae often have a long, thin spine (arista); and the wings have only one prominent cross-vein among a reduced set of horizontal veins. Males often have modified legs and wings, and the tip of the abdomen points down and forward. Many can be seen displaying and congregating on horizontal leaves, such as the mainly green species of *Sciapus*. Others are found on smooth-barked tree trunks. The adults and most known larvae are predators on other insects.

Hover Flies, Drone Flies (Family Syrphidae)

There are 169 species of these small to medium-sized, brightly coloured flies. They have a distinctive hovering flight, and are covered in dense, fine, often yellow, hair and generally mimic the warning colours of wasps and bees.

Native drone fly, *Eristalinus punctulatus*, Syrphidae (10 mm long)

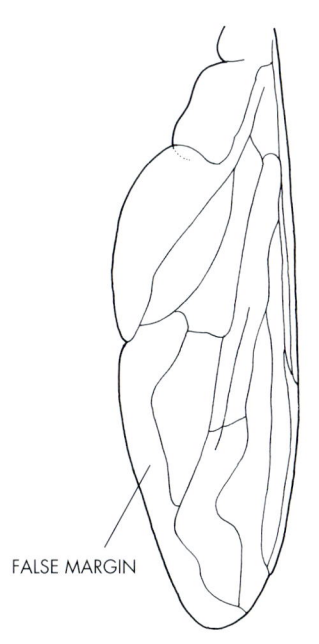

The main difference between this family (the hover flies) and the bee flies (Bombyliidae) is the hover flies' shorter wings, which are typically only as long as the abdomen and have a series of closed cells giving the appearance of a false margin (see Figure 46). Adults are most commonly seen near flowers and some produce a loud buzz similar to bees, e.g. the native drone fly, *Eristalinus punctulatus*. Syrphid larvae prey on aphids and feed on detritus and fungi; some are commensal with ants, and the well-named rat-tailed maggots live in polluted mud.

Figure 46. Wing of a typical hover fly (family Syrphidae) showing the false-margin effect created by veins which do not reach the margin. [R. Redman]

Antler Flies and Others (Family Platystomatidae)

These are small to medium-sized flies (236 species) which are similar in appearance to the fruit flies (Tephritidae). Wings are often mottled and coloured, with very similar venation to that of the fruit flies but with no sharply upturned Sc vein (see Figure 47). Adults can be commonly seen sitting on vegetation in wetter

FLIES

Ovipositor. The genus *Adrama* includes yellow species found in the tropics. This family contains many of the most important species of agricultural pests, including the Qld fruit fly (*Bactrocera tryoni*) and the introduced Mediterranean fruit fly (*Ceratitis capitata*). There is always the danger of other economically important fruit fly species entering Australia from Asia and therefore the family is well studied in Australia. Most larvae feed on fruit, but others use rotting wood, plant stems and the flower heads of daisies.

Vinegar Flies (Family Drosophilidae)

These are small flies (247 species) with diverse habits that are often attracted to rotting or fermenting fruit. They are especially numerous in tropical Qld

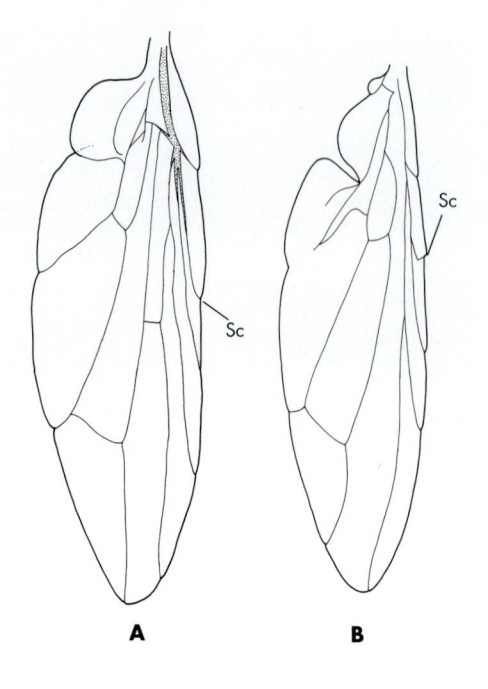

A   **B**

Figure 47. Wings of typical antler flies and fruit flies: (a) the Sc vein reaching the wing margin at a shallow angle (family Platystomatidae); (b) the Sc vein making a sharp turn up to the margin (family Tephritidae). [R. Redman]

areas, or feeding on fresh dung. Larvae utilise a variety of habitats including decaying wood and roots, and some are predacious on insect eggs. One interesting tropical genus is *Achias*, in which the females have a 'hammerhead' and the males have the eyes at the end of long stalks, or 'antlers', which they use in ritualised combat with other males.

Fruit Flies (Family Tephritidae)

There are 135 species of Tephritidae. They are small to medium-sized flies; many species are light brown with yellow markings. The wings are often patterned and the Sc vein is bent upwards, almost at a right angle (see Figure 47). When at rest, the wings are often rotated in a paddle-like movement and many perform courtship displays like the genus *Austronerva*. The female has the large

Antler fly, male, *Achias australis*, **Platystomatidae** (10 mm long)

ABOVE: Courting fruit flies, *Austronevra* sp., **Tephritidae** (6 mm long)

LEFT: Banana fruit fly, male, *Bactrocera musae*, **Tephritidae** (7 mm long)

RIGHT: Vinegar flies, *Drosophila* sp., **Drosophilidae** (3 mm long)

FLIES

Stilt-legged fly, *Mimegralla* sp., **Micropezidae** (12 mm long)

where clouds of tiny individuals may hover around the fruit bowl. The minute antennae have a thin plumose tip. The family is most noted for the genus *Drosophila* which is used worldwide in experiments studying genetic theory. Larvae are found in a variety of media, but fungi and yeast-infected vegetable matter are the most important.

Stilt-legged Flies (Family Micropezidae)

This small family of 18 species has a very distinctive shaped body raised

Neomyia sp., **Muscidae** (8 mm long)

Blowfly, *Amenia imperialis*, **Calliphoridae** (16 mm long)

high on long, thin legs; the head has a 'neck' and an abdomen with a 'waist'. Adults are predators.

House Flies, Bush Flies and Others (Family Muscidae)

This and the remaining three families are known as calyptrate flies as they have a calypter which is a lobe at the base of the wings that covers the halteres from above (see Figure 44b). On microscopic examination they also share a groove along the second antennal segment. Adults are usually stocky and strongly bristled. For typical wing venation see Figure 44b. There are 180 species in the family Muscidae. Two of the most familiar Australian flies belong to this family: *Musca domestica*, the cosmopolitan

FLIES

Australian sheep blowfly, *Lucilia cuprina*, Calliphoridae (8mm long)

house fly; and *M. vetustissima*, the bush fly. Both are a great nuisance, and the house fly is known to transmit important diseases such as typhoid. Some species of muscids are predators as adults; and some, like the stable fly, *Stomoxys calcitrans*, are blood sucking. The genus *Neomyia* is found in forests and feeds on dung. Muscid larvae live in many habitats, with carrion, dung, fungi and rotting vegetation being the most important.

Blowflies and Bluebottles (Family Calliphoridae)

There are 140 species of these small to medium-sized calyptrate flies. Many are metallic green or blue, others are mixtures of brown and black. Sets of bristles on the side of the thorax below the halteres distinguish this family and the Tachinidae from other calyptrate flies (see Figure 48), while the presence of a plumose arista on the antennae of calliphorids usually separates them from the tachinids. The larvae of most calliphorids develop in carrion or dung, although several small subfamilies are parasitic. There is usually a marked succession in the use of carrion by blowflies, with the later species often feeding on the larvae of the earlier species as well as the carrion. While many blowflies are beneficial by speeding up the decomposition of carcasses, there are some economically important pest species which attack living animals. The best known is the introduced *Lucilia cuprina* which causes 'blowfly-strike' in sheep when the larvae start feeding on inflamed skin under the wool, and spread into the flesh. The Oriental screwworm fly, *Chrysomya bezziana*, which infests cattle, has not yet found its way from Papua New Guinea into Australia. The genus *Amenia* contains some very colourful, widespread native species and species of *Stomorhina* can be found along beaches.

Flesh Flies (Family Sarcophagidae)

This group of 67 species is sometimes treated as a part of the Calliphoridae. They differ by not being metallic; by

Flesh fly, **Sarcophagidae** (10 mm long)

often having three longitudinal dark stripes on the top of the thorax; and by having only the first half of the antennal arista plumose. The majority breed in carrion or rotting vegetation, while some are parasites of other insects. The females are viviparous, producing live maggots directly onto the food source.

Family Tachinidae

This family contains 542 species. These are small to large flies (up to 35-mm wingspan), characterised by an enlarged subscutellum below and behind the scutellum (third division of the thorax—see Figure 48). The arista is usually not plumose, unlike the calliphorids. This is a very beneficial family as the larvae are all parasitic on other insects, including pest species in several orders. Many adults are grey and black, resembling muscids or sarcophagids. Others, such as the large *Rutilia* which are parasites of scarabacid beetles, have very showy metallic colours.

Tachinid, *Rutilia* sp., **Tachinidae** (20 mm long)

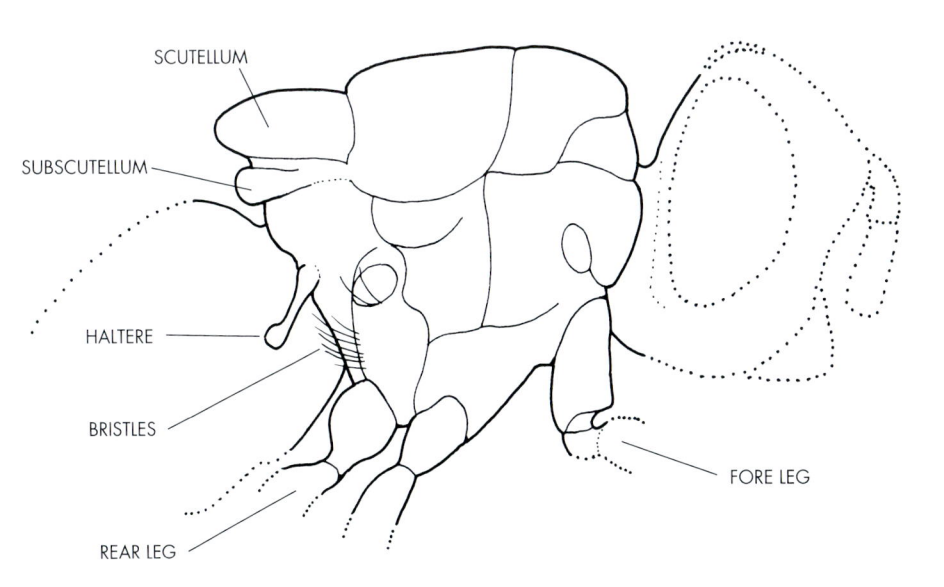

Figure 48. A side view of the thorax of a typical species of the family Tachinidae. Only tachinids have the well-developed subscutellum, while tachinids and calliphorids have the bristles below the haltere. [R. Redman]

Tachinid, *Senostoma* sp., **Tachinidae** (10 mm long)

Species of *Senostoma* have very long spindly legs and elongated mouthparts, and can often be seen on flowers.

Eggs are laid on or near the host. Those laid near the host may hatch to burrow in from outside, or be eaten by the host and hatch inside it. Those laid on the host can be on larvae or adult insects and it is common to see eggs attached to lepidopteran caterpillars. The larvae burrow inside and eat the host, usually resulting in its death.

Many species have the habit of not killing the host by the process of eating non-essential tissue. The parasites usually leave the host to pupate in the soil.

Further Reading

Colless, D. H. and McAlpine, D.K. (1991), 'Diptera' (ch. 39) in *The Insects of Australia*, CSIRO, Melbourne.

Oldroyd, H. (1964), *The Natural History of Flies*, Weidenfeld & Nicholson, London.

CADDISFLIES
ORDER TRICHOPTERA

World families: 43; species: 7000 **Australian families:** 25; species: 480

Characteristics of Order

Complete life cycle; small to medium (2–40 mm in body length); the elongate body and two pairs of wings are hairy; the forewings and hind wings are always of different size, but either pair may be the larger in different species; mouthparts are simple and reduced, with generally only the palps visible; antennae are usually filiform, tapering and half or more the body length.

The larvae are aquatic with well-developed legs and biting mandibulate mouthparts. They usually live in a case constructed of silk covered by various materials.

As the wings may sometimes appear to be covered in scales rather than just hair, there may be superficial confusion with moths (Lepidoptera). The main and simple difference is the mouthparts: moths have a prominent curled proboscis; caddisflies do not.

Life Cycle

The females lay up to several hundred eggs encased in a gelatinous mass on or near water. The larval thorax and head are sclerotised but the abdomen is very soft. Larvae are capable of producing silk from glands near the mouth and they use this to construct a variety of portable shelters, or cases, to protect their soft bodies and provide camouflage. The silk can also anchor them in strong currents and is sometimes used to create snares for prey. The larvae have a pair of anal prolegs which anchors them into the case and as they grow more material is added to the front end. The detritus used to coat the cases depends on the genus of caddisfly and its habitat and can include fine to coarse sand, plant parts and all manner of organic particles (Figure 49).

Pupation takes place underwater inside the larval case or within a specially made silken covering. The pupae have mandibles for chewing out of the case and then swim to a perch above the surface for the adult to emerge. Most of the life cycle is spent as a larva. The adults are generally very short lived.

Biology

Caddisflies are more familiar as larvae than as the adults. The larvae live in all types of fresh water habitats and a few even inhabit saline tidal areas. Water conditions such as temperature, oxygen content and other chemical and particle content are very impor-tant parameters for different species. This relationship allows the presence or absence of particular caddisfly species larvae to be used as an environmental quality indicator.

The larvae have many diets and means of feeding, including chewing

CADDISFLIES

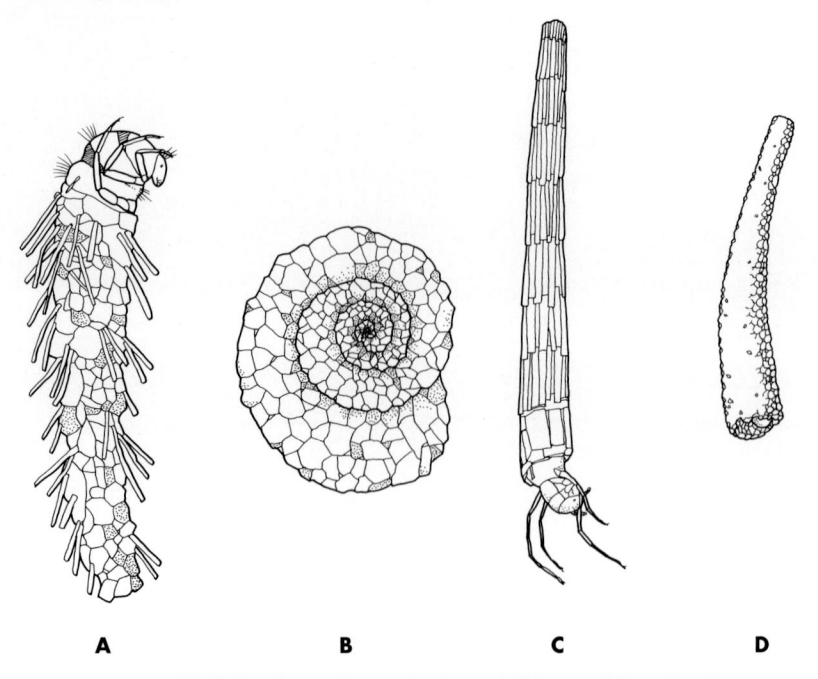

A B C D

Figure 49. Some caddisfly larval cases: (a) a marine larva of *Philanisus plebeius* (family Chathamiidae) uses sand and algae; (b) a larva of *Helicopsyche* (family Helicopsychidae) makes sand snail-shell imitation cases [after P. Tillyard]; (c) a larva of Leptoceridae often uses reeds; (d) a larva of *Merilia* (family Odontoceridae) lives in tapering tubes made of sand.

live or dead vascular plants and algae; filtering fine organic particles from the current or scraping them off rocks and plants; piercing plants to suck out the phloem or xylem; and catching mainly insect prey, sometimes in silken net traps. Some larvae are pests of ornamental aquatic plants like lilies and a few damage wooden parts of boats. In turn, their large numbers make them an important part in the diet of fish and many aquatic invertebrates. The adults tend not to feed, but are themselves seasonally eaten by many birds and bats.

Classification

Characteristics which distinguish the adults require a microscope and preparation of the wings to properly examine wing venation. Other features are the shape of the palps, various wart-like structures on the thorax and details of the genitalia.

However, the larvae are more likely to be encountered and while their body features can also be obscure, much can be inferred from the case materials and shapes. These tend to be consistent within individual genera rather than families and can be recognised with practice.

The Australian Trichoptera are placed by Neboiss (1991) into three superfamilies and 28 families. Some of the larger families and their features are discussed below.

Family Hydrobiosidae

The family Hydrobiosidae with 57 species, has moderate-sized adults (wingspan: 10–30 mm), with dark wings, often with mottled colour patterns. Antennae are as long as the forewings and ocelli are present. The larvae, which are predators in cool mountain streams, differ by not making cases.

Family Hydroptilidae

The family Hydroptilidae with 101 species, can be distinguished by having very small adults (wingspan: 4–12 mm), short antennae and very hairy wings. There is an extra fringe of hair along the hind wings that is longer than the width of the wing itself, and both pairs tend to be pointed rather than rounded. The larvae only make a case in the last instar, which is usually purse shaped and covered in sand, detritus or silk only.

Family Calamoceratidae

Calamoceratidae with only 10 species in the genus *Anisocentropus* are well-defined by their sub-triangular forewings and very hairy and large maxillary palps. The larvae live in a flat case made of two pieces of leaf.

Family Stenopsychidae

Stenopsychidae with nine species in the genus *Stenopsychodes* has some of the most striking caddisflies in Australia. They are large (wingspan: 18–35 mm) and have noticeably thick antennae which are the length of the body.

Family Ecnomidae

Ecnomidae with 57 species, has small to moderate-sized adults (wingspan: 6–18 mm). The wings tend to be dull grey; the antennae are less than the length of the forewing; and ocelli are absent. The larvae construct open-ended tubes which are coated with a variety of local debris including plant pieces and sand.

Family Leptoceridae

Leptoceridae with 83 species has medium to large adults (wingspan: 10–40 mm). They are distinctive by having antennae which are twice as long as, or longer than their very slender forewings. In many, the wing colour is dark and the antennae light, further emphasising the family features. The larvae which can be detritivors or predators, construct various tubular-style cases, often tapering at the end and using many types of coating. *Triplectides* is a typical east coast genus. Larvae of the genus

Anisocentropus banghasi, **Calamoceratidae** (15 mm long)

Stenopsychodes sp., **Stenopsychidae** (18 mm long)

Longhorned caddis, *Triplectides* sp., **Leptoceridae** (15 mm long)

Oecetis make a home out of bits of plants which they arrange like planks in a log cabin.

Family Helicopsychidae

The small family Helicopsychidae has unremarkable adults, but the larvae are distinctive by making sand cases in the shape of snail shells (see Figure 49). The remaining 21 families each have few species and contain about 50 Australian genera among them.

Further Reading

Neboiss, A. (1982), 'The caddisflies (Trichoptera) of south-western Australia', *Australian Journal of Zoology*, vol. 30, pp. 271–325.

Neboiss, A. (1991), 'Trichoptera' (ch. 40) in *The Insects of Australia*, CSIRO, Melbourne.

MOTHS AND BUTTERFLIES
ORDER LEPIDOPTERA

World families: 127; species: > 100 000 **Australian families:** 82; species: 20 816

Characteristics of Order

Complete life cycle; small to very large (wingspan: 3–250 mm); wings and body are usually covered by minute overlapping scales; in most species, mouthparts are in the form of a coiled tube (proboscis), with the labial and maxillary palps retained alongside; large compound eyes, sometimes hairy; antennae can vary from simple filiform to fanlike in males of some families and clubbed in butterflies; legs are usually covered in scales and hind tibia have two pairs of spurs.

The larva is a typical caterpillar, with a sclerotised head and soft thorax, and usually 10 visible abdominal segments; three pairs of legs on thorax are visible and the abdomen has short prolegs on the third to sixth, and tenth segments. The important difference between these and other caterpillar-like larvae (such as the sawfly wasps, Hymenoptera) is the presence of a ring of fine hooks (crochets) on the ends of these prolegs.

Life Cycle

This order is often used to demonstrate the classic egg–larvae–pupa–adult life cycle (the endopterygote cycle). Butterflies tend to find their mates by sight, but most moths depend on the males finding the females by a scent (pheromone) that the females release. Eggs are usually laid singly or in groups on the larval food plant, numbers varying from about 50 to several thousand.

The majority of families have caterpillars that live externally on the foliage of plants, but some are miners within leaves, some burrow in stems and others live in roots. All can produce silk which is usually only employed as a lifeline when they fall, but in some groups is used to construct complex shelters and to spin a case or cocoon for the pupa. The average number of larval instars is four or five, and the larval period may last as little as a few weeks in many exposed leaf feeders, to a year or more in some wood-tunnelling species, and up to five years in the giant wood moths of the family Cossidae.

To pupate, many moth larvae find a sheltered spot and spin a silken cocoon. Species that live in wood make an exit hole which they then cover with frass and pupate inside the wood. Others make underground chambers where they pupate with minimum silk protection. Butterflies other than skippers have an exposed pupa that usually hangs upside down from vegetation. These are known as a chrysalis.

Biology

The vast majority of caterpillars are herbivores, either eating foliage or wood. They feed day and night and can be major defoliators which makes many species serious pests of economically important crops. However, a small number feed on soft insects like scale insects, others on ant larvae, and cannibalism is common in many otherwise herbivorous families. Adult lepidopterans mainly feed on nectar from flowers which is good fuel for flight but is also important for egg development. Some adults have reduced mouthparts and live only on reserves stored by the larvae.

Day-flying moths and butterflies are often very brightly coloured which, in some species, demonstrates an unpalatable taste to predators. In many, this is the result of eating poisonous plants as larvae and storing some of the toxins. In others that are palatable, the colour represents mimicry of warning colours which include red, black and yellow patterns in many insect orders. Many caterpillars living exposed on leaves also employ such colours and may add to this by a discharge of foul fluids, sudden movement, irritating hairs and even specialised poisonous tufts.

Nevertheless, both adults and larvae are an important part of the diet of many other insects, birds and especially bats. In some species the relationship with bats includes the development of hearing organs which detect the presence of bat ultrasound. In some, this initiates wild evasive manoeuvres or a straight plummet manoeuvre. The pinnacle of this co-evolutionary development is among the unpalatable Arctiidae moths, which produce an ultrasound call that bats learn to associate with their bad taste, just as birds learn the association of colours and taste in day-flying species.

Classification

Recent classification (Nielsen & Common 1991) recognises four suborders. However, three of these are very small with only 10 species in Australia, and so in this brief treatment of the major families of an order with over 20 000 species they will be disregarded. Family characteristics tend strongly towards wing venation, and aspects of the palps and antennae (see Figure 46). In moths, generic characteristics tend to be easier to pick than family ones and therefore the following text will include only main visible family features, though these rarely represent a foolproof guide to a whole family. An important diagnostic feature, sadly lost with pinned specimens, is the posture of live insects. Such features, where of help, are included in the text and demonstrated in pictures.

The Butterfly–Moth Question

What is the difference between butterflies and moths? Butterflies are simply several day-flying families of moths. They have clubbed antennae, sit with their wings upright and lack a wing-coupling spine along the hind wing called a frenulum. There are a small number of other day-flying moths with clubbed antennae, but no moths with all these characteristics in combination. Australia has about 400 species of butterflies in the superfamilies Hesperoidea (skippers) and Papilionoidea.

Following is a look at the major families of moths and butterflies, and some smaller ones.

Moths

Superfamily Hepialoidea

The superfamily Hepialoidea has four families characterised by the 150 species of ghost moths and swift moths in the Hepialidae. Many have the unique habit of hanging from

Splendid ghost moth, *Aenetus ligniveren*, **Hepialidae** (40 mm long) [J. Landy]

MOTHS AND BUTTERFLIES

Moerarchis australasiella, **Tineidae** (15 mm long)

branches by their forelegs. The larvae are mainly tunnellers in the stems or roots of living trees and sometimes in the ground. The adults include some of the handsomest insects, easily rivalling the butterflies. The genus *Aenetus* has many large species (up to 150-mm wingspan) that range from greens to deep gold and even blue.

Superfamily Tineoidea
This superfamily contains eight families with 1385 species.

Clothes Moths (Family Tineidae)
The namesake family, Tineidae, has 440 species which include the cosmopolitan pest, the clothes moth, *Tineola bisselliella*. Many of the native species are also associated with animal fibre and some with stored grain and fungi. Most adults are small with wingspans rarely over 25 mm. The wings are long and narrow, often with a silvery sheen and a hair fringe along the hind margins. The head is adorned by a dense mop of

Case-moth caterpillar, **Psychidae** (30 mm long)

RIGHT:
Wingia aurata,
Oecophoridae
(16 mm long)

BELOW RIGHT:
Pseudaegeria sp.,
Oecophoridae
(14 mm long)

hair, as in the large east coast *Moerarchis australasiella*. Adults have reduced mouthparts and do not feed.

Family Gracillariidae

The 450 species of the Gracillariidae are very similar to tineids, but can be distinguished by their habit of sitting up on their long front legs. Their larvae are mainly leaf miners.

Case Moths (Family Psychidae)

Another well-known family in this group is the Psychidae, with 350 species of case moths or bag moths. In most species, the caterpillar lives in a mobile case made of silk and various combinations of leaves and sticks, and feeds on foliage. The cases can be diagnostic of the species. The adult males have robust bodies, often feathery antennae and wings without obvious fringes. They tend to fly very erratically in search of the females, which generally remain in the cases and do not develop wings.

Superfamily Gelechioidea

The superfamily Gelechioidea has 8690 species in 13 families. The main diagnostics for this large group are the proboscis which is clothed in scales, and the labial palps which are also covered in scales and curved up and back like tusks.

Family Oecophoridae

The major family in this superfamily is the Oecophoridae with 5550 species. A very subtle and not universal characteristic of this family is a tuft of hair at the base of the antennae (see Figure 50). The larvae are unusual in being specialised feeders on mainly dead leaves, and their huge variety and number therefore makes them important in nutrient recycling. Adult wingspans are generally about 20 to 40 mm, though they can reach 70 mm.

Macrobathra sp., **Cosmopterigidae** (10 mm long)

Morphotica mirifica, **Cosmopterigidae** (9 mm long)

A fringe is prominent around the hind wing, and the antennae tend to be around two-thirds of the forewing length. The genus *Wingia* has widely distributed species feeding on eucalypts, while members of the genus *Pseudaegeria* are diurnal and show warning colouration.

Family Gelechiidae

The Gelechiidae, with 1580 species, is the next largest family. They are generally smaller, with wingspans from 7 to 32 mm, and the antennae are just short of the forewing length. The larvae feed on living leaves and most species live in characteristic shelters of two leaves joined together with silk.

Family Cosmopterigidae

The Cosmopterigidae, with 850 species, have very narrow wings; the hind wings are often only a thin line with a strong hair margin, looking like feathers. They are smaller still, with wingspans from 6 to 32 mm, but usually under 20 mm. The genus *Macrobathra* has over 100 species with bold markings on a dark background. Species of *Morphotica* have very prominent palps, displaying the curve typical of this superfamily. The larvae have broad feeding habits from foliage and fungi, to wood, with some even being predators of scale insects.

Superfamily Cossoidea

Goat Moths (Family Cossidae)

The small superfamily Cossoidea contains the wood moths or goat moths in the Cossidae. The 200 species have large, fat bodies, the heaviest of all moths, and vary in size from 10- to 240-mm wingspans. The proboscis is so reduced as not to be visible in some, as the adults do not feed. The larvae are wood borers and are the original witchetty grub of central Australian Aboriginal peoples. Larvae may take several years to complete development. Adult colours are dominated by grey, like the tropical *Endoxyla mackeri*, with a 170-mm wingspan.

Superfamily Tortricoidea

Family Tortricidae

The superfamily Tortricoidea has one large leafroller family, the Tortricidae, with 1230 species. A general characteristic is the rather squared ends of the wings, the forewings tending to be rectangular shaped. Wingspans range from 8 to 35 mm, with most under 25 mm. Antennae are generally half the forewing length, and the labial palps are pointed forward and scaled

MOTHS AND BUTTERFLIES

(see Figure 50). The majority of adults have cryptic colours and hide among leaf litter or on trees. *Ophiorrhobda phaesigma* is a showy tropical species. The larvae mainly feed on living and dead leaves and are usually hidden in folded or silk-joined leaves.

Superfamily Zygaenoidea

The small Zygaenoidea has 227 species in four families, two of which are likely to be encountered.

Cup Moths (Family Limacodidae)

The Limacodidae with 115 species are the cup moths, more famous for their caterpillars than the adults. These squat, slug-like larvae have projections at the front and back that often bear a whirl of stinging spines that look very much like minute sea anemones. Their colours are often striking contrasts added to a basic green to warn predators. Adults have squat, fat and furry bodies with generally broad wings up to 70 mm across.

Wood moth, *Endoxyla mackeri*, **Cossidae** (80 mm long)

Leafroller moth, *Ophiorrhobda phaesigma*, **Tortricidae** (9 mm long)

MOTHS AND BUTTERFLIES

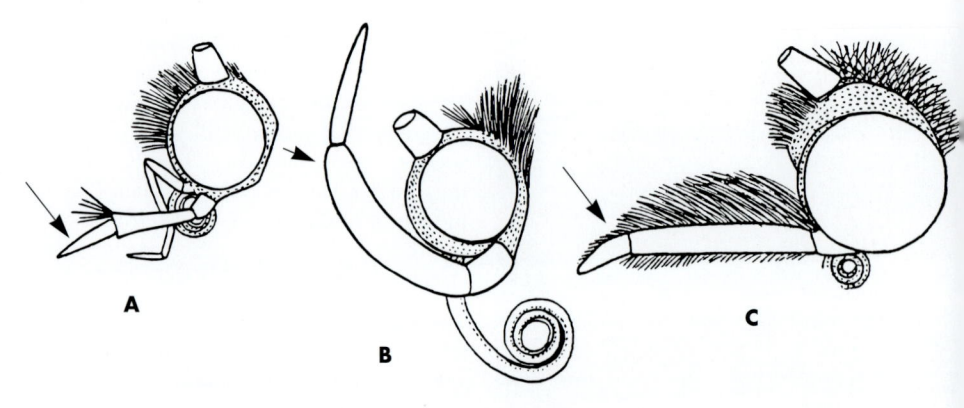

Figure 50. Profiles of the heads of major moth families showing the diagnostic position, shape and size of the labial palps (arrow) (the antennal position is illustrated by the first segment only on top): (a) Tineidae, note mop of 'hair' forward of the antennae and prominent other (maxillary) palps; (b) Oecophoridae with 'tusk-like' palps; (c) Tortricidae with large forward pointing palps. [R.J. Tillyard]

Burnets (Family Zygaenidae)

The family Zygaenidae, with 56 species, includes many day-flying species called burnets. They have very bright, even metallic colours dominated by blue, green, orange and black. Wingspans range from 12 to 110 mm, and their antennae are often thickened and sometimes clubbed. The larvae are similar to the cup moths but not as broad, are usually without stinging hairs and, in contrast to the adults, are nocturnal. The large genus *Lactura* is mainly tropical and adults often come to lights.

Cup moth caterpillar, **Limacodidae** (16 mm long)

MOTHS AND BUTTERFLIES

Cup moth, *Comana mittogramma*, **Limacodidae** (18 mm long)

Burnet, *Lactura erythractis*, **Zygaenidae** (15 mm long)

MOTHS AND BUTTERFLIES

Superfamily Pterophoroidea
Plume Moths (Family Pterophoridae)

The Pterophoroidea contains the small but very distinctive plume moth family Pterophoridae. The 40 species are small and very delicate (wingspan: 9–30 mm). What immediately distinguishes them is the division of the hind wing into three feather-like parts and the forewing into two plumed ends.

Plume moth, *Stangeia* sp., **Pterophoridae** (14 mm across)

Superfamily Pyraloidea
Family Pyralidae

The Pyraloidea contains the 1670 species of Pyralidae. This is the only other family (together with the Gelechioidea, p. 159) which has a scale-covered proboscis. The main visible, though often subtle, different characteristic is the labial palps. Unlike the recurving tusk-like ones of the gelechiids, pyralids usually have straight beak-like, often large palps. A less visible character is the presence of tympanal organs (ears) at the base of the abdomen. In general, pyralid wings differ by a more triangular shape of the forewings, broader hind wings and in posture. Gelechioids tend to fold their wings narrowly over and even around their bodies; while pyralids tend to sit with their wings more open in a distinct triangle, and often sit up at the front on very long legs. This combined posture is the strongest family characteristic. The larvae have many different habits including most ways of life on and in plants. One group is aquatic, whose larvae possess gills and build portable leaf shelters similar to those of caddisfly larvae.

Pyralid, *Aetholix flavibasalis*, **Pyralidae** (18 mm across)

Superfamily Geometroidea

The Geometroidea contains the second largest moth family, the Geometridae, sometimes known as emerald moths, with 2310 species.

Family Geometridae

They range in size, with wingspans from 12 to 120 mm, but tend to have a uniform general shape and stance. The forewings are broad and roughly triangular, and the hind wings are nearly as broad. The sitting posture involves holding the wings spread out flat against the substrate, usually with the top of the forewings forming an almost straight line at right angles to the body. Many adults are green, and colours tend to be mottled which, combined with the flat, shadowless posture, makes them very cryptic.

Pyralid, *Glyphodes canthusalis*, **Pyralidae** (25 mm across)

Pyralid, *Pygospila* sp., **Pyralidae** (40 mm across)

LEFT: Looper,
Problepsis apollinaria,
Geometridae
(35 mm across)

BELOW: Looper,
Pingasa chlora,
Geometridae
(35 mm across)

RIGHT: Looper,
Aeolochroma turneri,
Geometridae
(30 mm across)

ABOVE: Looper, *Anisozyga pieroides*, **Geometridae** (30 mm across)

LEFT: Looper, *Comostola chloragyra*, **Geometridae** (18 mm across)

RIGHT: Looper, *Alloeopage cinerea*, **Geometridae** (35 mm across)

MOTHS AND BUTTERFLIES

Looper caterpillar, *Dysphania fenestrata*, **Geometridae** (35 mm long)

The larvae are usually known as 'loopers'. They are thin and somewhat cylindrical and progress by stretching their front half to a new position and then bringing up the rear in one action —their other name, derived from this motion, is inch-worm. When motionless, some species can stay erect and mimic twigs. Other species are flattened and attach local frass to their bodies for excellent concealment.

Superfamily Bombycoidea

The superfamily Bombycoidea has 245 generally stout species in six families. They have fat, woolly bodies with broad wings and feathered (pectinate) antennae. The proboscis is very reduced or absent; and the maxillary and even the labial palps are short.

Family Saturniidae

Members of the emperor moth family, Saturniidae, are large (wingspan:

Looper caterpillar, **Geometridae** (15 mm long)

RIGHT: Hercules moth caterpillar, *Coscinocera hercules*, **Saturniidae** (100 mm long)

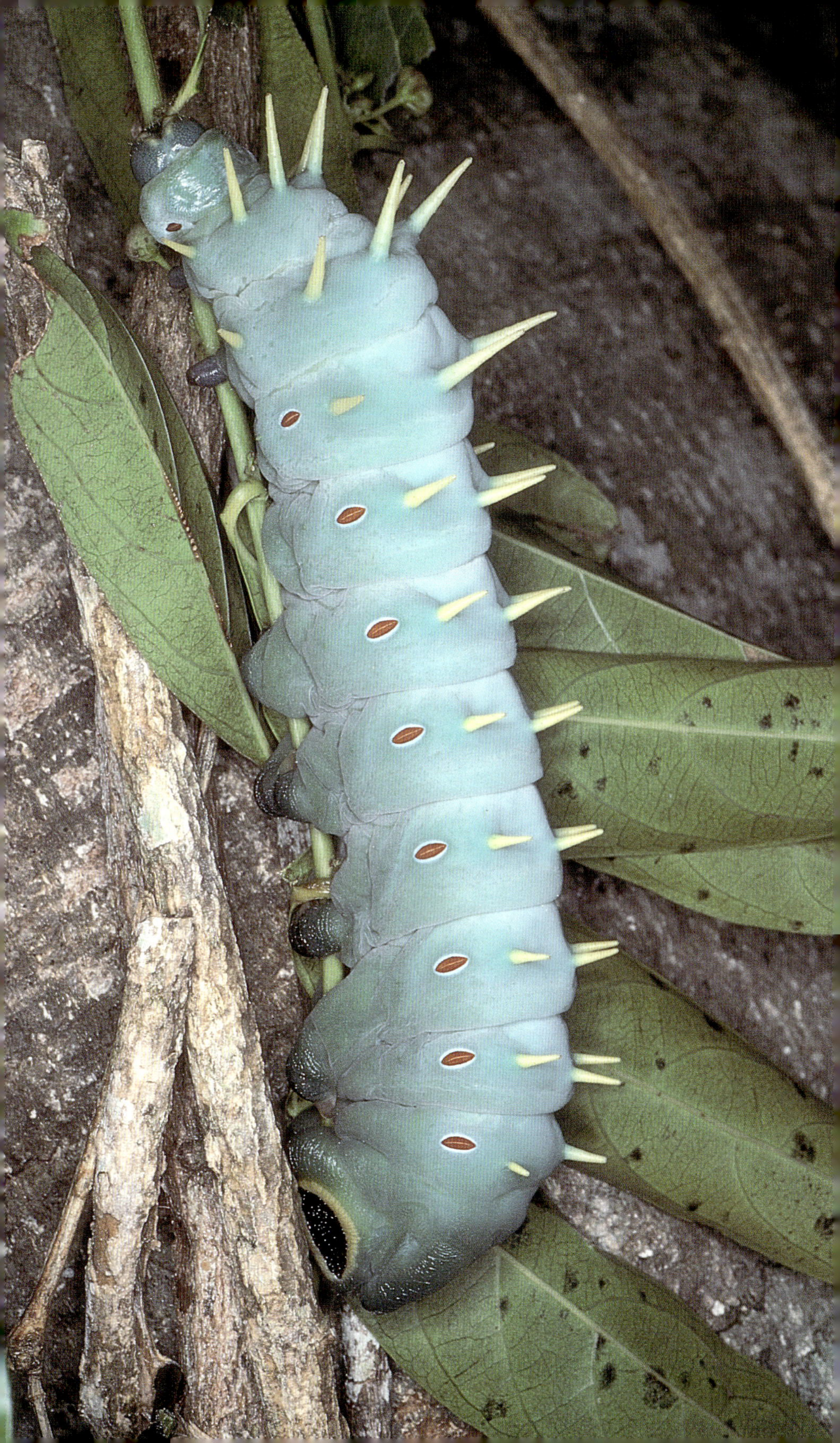

Emperor gum moth, *Opodiphthera eucalypti*, **Saturniidae** (75 mm across)

Anthela sp., **Anthelidae** (50 mm across)

65–270 mm) and include one of the largest moths in the world, the tailed Hercules moth of north Qld, *Coscinocera hercules*, which has a 100-mm long caterpillar. The most widespread species is the emperor gum moth *Opodiphthera eucalypti* found from the NT to Vic.

Family Anthelidae

The major family is the Anthelidae with 121 small to large species (wingspans: 25–160 mm), found only in Australia and New Guinea. The short labial palps are forward pointing and beak-like. Nearly half the species are in the genus *Anthela* with rich yellow, brown or green hues.

Superfamily Sphingoidea
Hawk Moths (Family Sphingidae)

The superfamily Sphingoidea has the well-known, mainly tropical hawk moths in the family Sphingidae. Only 65 species are found in Australia, and these often large (wingspan: 40–190 mm) and handsome moths are sought after by collectors. They are very robust of body and are strong fliers, with a shape more obviously aerodynamic than most insects. The proboscis is prominent and often very long for inserting into flowers while the moth hovers. The large caterpillars have a prominent spine arising from the tail and are often very colourful. The typically shaped genus *Hippotion* has five species and *H. scrofa* is found

Scrofa hawk moth, *Hippotion scrofa*, **Sphingidae** (40 mm long)

throughout Australia on a wide variety of food plants. The large *Ambylux wildei* is a rainforest species in Qld.

Superfamily Noctuoidea
Family Noctuidae

The Noctuoidea is a large and very diverse group of seven families with 2676 species. The major family is the Noctuidae with almost 2000 species and a sufficient variety of shapes and stances to be very hard to define. A subtle characteristic that defines this family is the presence of tympanal organs (ears) at the back of the thorax. Only the Geometridae (a distinctly different family) also have 'ears' there (see p.165). In many noctuids, ocelli are visible through the scales on the head. They are small to large species (wingspan: 10–170 mm), divided into 20 subfamilies, and include some well-known crop pests such as the

Hawk moth, *Ambylux wildei*, **Sphingidae** (90 mm across)

MOTHS AND BUTTERFLIES

Budworm caterpillar, *Helicoverpa armigera*, **Noctuidae** (20 mm long)

Speiredonia zamis, **Noctuidae** (55 mm across)

MOTHS AND BUTTERFLIES

Periopta ardescens, **Noctuidae** (25 mm across)

Amerila rubripes, **Arctiidae** (30 mm long)

budworms, *Helicoverpa*. *H. armigera* attacks a wide variety of crops and grasses, and through its migratory skills is a good opportunist that is hard to control. Migration is a feature of this family, and something the bogong moth *Agrotis infusa* is very famous for. The population passes Canberra on its annual route and often takes over the city lights. One subfamily, Agaristinae, are brightly coloured day-fliers. *Periopta ardescens* is an east coast species. Many species in this family have 'eye spots' that may be used as a startle display when threatened. *Speiredonia zamis* is an example from the tropics. Generally, noctuid caterpillars have distinct segments, formations of tufts of short bristles, and bright colours. They tend to live exposed on foliage and their voracious appetites have given some species the names armyworms and cutworms.

Tiger Moths (Family Arctiidae)

The Arctiidae, with 342 species, are the very distinctive tiger moths. Many have warning (aposematic) colouration and some are day-flying. Day-fliers advertise their distaste to predators by colour, and some night-fliers can warn bats by producing ultrasonic sound. The genus *Amerila* even produces a foul chemical froth when threatened. The genus *Amata* has 36 species utilising the common

173

MOTHS AND BUTTERFLIES

Amata marella, **Arctiidae** (30 mm across)

Tussock moth, *Lymantria antennata*, **Lymantriidae** (25 mm long)

pattern of black wings with yellow to white dots, and striped abdomens, e.g. *Amata marella*, an east coast species.

Family Lymantriidae

Another distinctive family in this group is the Lymantriidae (the tussock moths) with 129 very furry species. Most have squat, fat bodies with the scales very hair-like, especially between the wings and spreading onto their inner margins. They are small to large (wingspan: 15–100 mm), with feathery antennae and no proboscis. The caterpillars are also very hairy. *Lymantria antennata* is a large species found in Qld.

Butterflies

The butterflies are currently included in two superfamilies. Treatment here is cursory, as there are several excellent guide books on Australian butterflies available to consult.

Superfamily Hesperioidea
Skippers (Family Hesperiidae)

The Hesperioidea contains the skippers in the family Hesperiidae. They differ from other butterflies by their broad heads with the antennae spread widely apart and thickening towards the end, but still coming to a curved point rather than being truly clubbed. Most of the 122 species are small (wingspan: 20–50 mm), with brown and yellow being the dominant colours. Skippers usually sit with wings held up and together, like other butterflies, or in a unique stance with the forewings up and the hind wings spread like the species of *Ocybadistes* from east coast grasslands. The caterpillars feed mainly on grasses within rolled leaves.

Skipper, *Ocybadistes* sp., **Hesperiidae** (10 mm long)

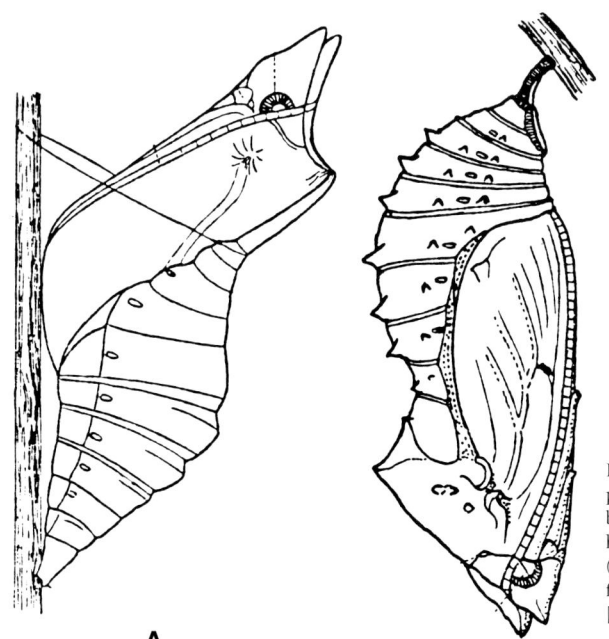

Figure 51. Two types of pupae (chrysalises) found in butterflies: (a) Papilionidae, held upright by a silk belt; (b) Nymphalidae, hanging free, upside down. [R.J. Tillyard]

Superfamily Papilionoidea

The Papilionoidea, with 275 species in four families, contains the rest of the butterflies.

The pupae can hang either by the base alone, or with a silk 'safety belt' holding them upright against a stem, or be simply in the ground (see Figure 51).

Swallowtails (Family Papilionidae)

The swallowtails have only 18 Australian species. These include the triangles (six species of *Graphium*); the orchard butterfly, *Papilio aegeus*; the vivid blue Ulysses, *Papilio ulysses*; and the giant birdwing, *Troides priamus* with a wingspan of up to 140 mm. The caterpillars are stouter than those of most other butterflies and have a forked process that can be everted from the top of the thorax to produce a defensive odour. Pupae are always held upright by a silk belt.

Whites (Family Pieridae)

The whites have 32 species that are mainly yellow and white with various patterns around the perimeter of the wings. The caterpillars are similar to many moth families (e.g. the Noctuidae), with patches of short stout hairs. The pupae can be hanging or upright. The most famous species is the cabbage white, *Pieris rapae*, an introduced pest. Among the native species are the jezebels (eight species of *Delias*) with conspicuous red and yellow underwings; the grass yellows (seven species of *Eurema*); the pearl whites (four species of *Elodina*); and the migratory caper white, *Belenois java*.

Browns (Family Nymphalidae)

The browns and fritillaries include 85 Australian species. Most have combinations of browns on their wings, sometimes with concentric spots like

Male birdwing, *Troides priamus*, **Papilionidae** (140 mm wingspan)

ABOVE: Common grass yellow, *Eurema hecabe*, **Pieridae** (25 mm wingspan)

LEFT: Evening brown, *Melanitis leda*, **Nymphalidae** (50 mm wingspan)

Small dusky blue, *Candalides erinus*, **Lycaenidae** (25 mm long)

Melanitis leda which feeds on grasses in eastern Australia. Their caterpillars tend to have long paired filaments, resembling anemone tentacles, on top of several body segments and concentric stripe colouration. The pupae always hang free. The most famous species is the worldwide migrating wanderer butterfly, *Danaus plexippus*. Its brick-red colour is a warning of the poisons butterflies of this genus store from eating milkweed (*Asclepias* sp.) as larvae.

Blues (Family Lycaenidae)

The final butterfly family is the blues with 140 Australian species. While many colours are present on the underside of the wings (as seen in living species at rest), the topside wing colours (as seen at takeoff and in flight) are mostly based on blue, sometimes purple, and rarely red. The larvae are flattened, usually green, and often attended by ants on the foliage they feed on. The large *Liphyra brassolis*, the ant butterfly, takes this association further by having a caterpillar which lives inside the nests of green tree ants and eats the ant larvae. The most colourful species belong to the large genera *Hypochrysops* and *Ogyris*. A typical and widespread genus is *Candalides*, with 12 pale blue and grey-white species.

Further Reading

Common, I.F.B. (1990), *Moths of Australia*, Melbourne University Press, Vic.

Common, I.F.B. and Waterhouse, D.F. (1981), *Butterflies of Australia*, 2nd edn, Angus and Robertson, Sydney.

Nielsen, E.S. and Common, I.F.B. (1991), 'Lepidoptera' (ch. 41) in *The Insects of Australia*, CSIRO, Melbourne.

WASPS, ANTS, BEES AND SAWFLIES
ORDER HYMENOPTERA

World families: 91; species: > 100 000 **Australian families:** 71; species: 14 781

Characteristics of Order

Complete life cycle; minute to large (0.15–120 mm in body length); the body usually has a constriction, or 'waist', between the first two abdominal segments, seemingly separating the thorax from the abdomen; two pairs of membranous wings usually with reduced to very reduced venation, and forewings are always longer than hind wings; head with mandibulate mouthparts, large compound eyes and usually three ocelli; short to long antennae which are strongly elbowed in some groups; prominent ovipositor in many groups which is modified to form a sting in some.

The larvae are maggot-like, legless, with discernible mandibles in the majority. The suborder Symphyta has caterpillar-shaped larvae with legs and often prolegs, though without the crochets of lepidopteran prolegs (see Figure 52).

Life Cycle

This order has several types of life cycle which will be discussed in greater detail under the relevant families. In many species, females can produce offspring without mating. The females that mate can store sperm and therefore can spread out egg laying to coincide with the availability of food, and even have control over the sex of the offspring, which has an effect on population growth. Primitive wasps such as the sawflies (Symphyta) lay eggs on vegetation because their larvae are leaf feeders. Other groups have leaf-mining larvae, and some families induce host plants to form galls within which the larvae develop.

The majority practise some form of parasitism, usually of other insects including other wasps, and also of spiders. Hosts can be the eggs, the immature stages and, more rarely, adult insects. Some families practise hyperparasitism, where another wasp parasite is in turn parasitised. Interestingly, this sometimes involves the gamble of laying eggs on a host that has not yet been parasitised by the wasp the hyperparasite is waiting for. There are species that are specific to one host and others that parasitise a variety of hosts.

The other important hymenopteran life style is that of the predatory, solitary wasps. These provide their larvae with dead or paralysed insects sealed in burrows and chambers where larval to adult development takes place in comparative safety. A variation of this is practised by the social wasps, ants and bees which form colonies where special castes provide for the larvae in large communal nests. The ants and wasps usually provide their larvae with insect and sometimes vegetarian food items, while the bees provide nectar and pollen.

Hymenopteran pupae are often enclosed in a thin, spun cocoon which

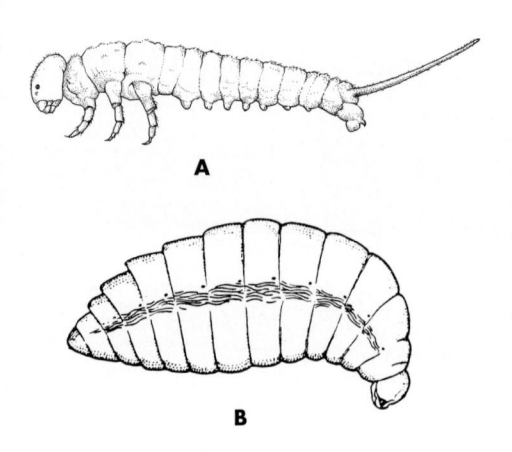

Figure 52. The main wasp larval types: (a) the sawfly (suborder Symphyta) which has legs and prolegs; (b) a typical predatory wasp larva (family Vespidae, suborder Apocrita) which has no legs. [R. Tillyard]

usually remains in or near the food source. Life cycles can be very brief in egg parasites, and longer in social species.

Biology

Adult wasps usually feed on nectar and therefore are often seen at flowers. Most species are diurnal and spend much of their time searching either for hosts to parasitise or prey to immobilise and feed to their larvae. Many parasitic species search for hosts by locating the food plants of the hosts and then use sense organs on their antennae to locate the actual prey. Some predatory wasps tend to dig chambers for the paralysed prey on which they will lay their eggs. These species use a very well-developed spatial memory to find the prepared nest sites after catching their prey. Others construct nests from mud, and these species are often seen on the edge of water, collecting moisture for the mix. Different nest types include the paper nests of the eusocial wasps (Vespidae) which also need water in the process of creating the pulp for the paper.

The true social wasps, bees and most ants have a complex caste system with a division of labour which usually has some workers caring for the offspring, while others build and defend the nest, and hunt and gather food. Usually only one queen per colony is in a reproductive state, and other queens and a male caste develop only when conditions are right to disperse to found new colonies. Among the bees there are also social parasitic species which invade the nests of other bees and induce the workers to care for a new queen and her progeny.

Classification

The Australian Hymenoptera were placed by Naumann (1991) in 71 families within two suborders, making it the third largest order after Coleoptera (beetles) and Lepidoptera (moths). The main characteristics used in the taxonomy of Hymenoptera are antennal and other head features; various thoracic characteristics, including the legs, genitalia and the wing venation. As in the chapter on Diptera (flies), it is not possible to present the major families without resorting to showing some wing venation features which define some families.

Most Hymenoptera are in the suborder Apocrita. Only the sawflies are in the small suborder Symphyta. The differences between the two are described below.

Sawflies (Suborder Symphyta)

This small group is poorly developed in Australia with only 176 species in six families. They are considered the most primitive families in the order, and differ from the true wasps, bees and ants by the following features. The main characteristic is the lack of a strong division, or 'waist', between the thorax and abdomen (which is created by the second abdominal segment

WASPS, ANTS, BEES AND SAWFLIES

Spitfire grubs, *Perga* sp., **Pergidae** (15 mm long)

forming a constriction or petiole). The wings have more complex venation resulting in more complete cells; the thorax has special pads on top (cenchri), which hold the wings in place at rest; there is a saw-like tip to the female ovipositor; and the larvae usually have a well-developed head and legs, and also have at least six pairs of abdominal prolegs. (The larvae are very similar to caterpillars of Lepidoptera, but can be distinguished by often having six or more pairs of prolegs, while lepidopteran caterpillars never have more than five pairs.) Sawflies are small to large (3–55 mm in body length), do not sting, and are often seen on flowers feeding on nectar. Most larvae are phytophagous, usually on the surface of leaves, though some bore in wood and stems, or mine leaves. One family only (Orussidae) is parasitic, mainly on beetle larvae.

Sawflies (Family Pergidae)

The largest family is the Pergidae, with 140 species. Most larval pergids feed on native trees and shrubs, are caterpillar-like and often sit together in tight formations. The defensive ability of some species to 'spit' out

Sawfly, *Platypsectra analis*, **Pergidae** (13 mm long)

181

Hatchet wasp, **Evaniidae** (9 mm long)

unpleasant liquids has given some the name 'spitfires'. *Perga* is the dominant genus, often defoliating young eucalypt trees in southeastern Australia. The genus *Platypsectra* has feathery antennae and strong colours warning that it is unpalatable.

Family Siricidae

The introduced European family Siricidae contains a major pest of pine plantations, *Sirex noctilio*, whose larvae feed inside *Pinus radiata*, and can kill the tree by introducing a toxic fungus.

Wasps, Ants and Bees (Suborder Apocrita)

The remaining Australian Hymenoptera are placed in Apocrita within 14 superfamilies and 65 families. The abdomen of Apocrita is divided into two parts. The first segment is attached to the thorax (making the thorax appear to have four rather than the usual three divisions), and is called the propodeum. The second segment is usually constricted and often elongated to form a 'waist' or petiole, behind which the remaining abdominal segments form the gaster.

Apocrita also lack the cenchri pads; the ovipositor is not serrated; the wing venation is often very reduced from the sawfly pattern; and the larvae do not have legs and often have a reduced head. The Apocrita is sometimes further divided into two groups. The Terebrantia (or Parasitica) includes mostly small species with a parasitic life cycle, no sting and no complex social organisation. The Aculeata includes the usually larger wasps, bees and ants that have the ovipositor modified into a sting for injecting poisonous and paralysing liquids, and some social species. Some of the major and/or distinctive families or superfamilies are discussed below.

(a) Suborder Apocrita— Terebrant (Parasitica) Group

Hatchet Wasps (Family Evaniidae)

A small family (40 species) of small to medium wasps (2–15 mm in body length). The thorax is stout and rounded, while the gaster (abdomen) is laterally flattened and attached from the very top of the propodeum by a long, narrow waist, looking like a flag or hatchet. As the adults walk they wave this 'flag' up and down. Most species are coloured from dark red to black and the body is finely punctured. The larvae are all parasitic on the egg cases of cockroaches.

Ichneumon Wasps (Family Ichneumonidae)

This is one of the largest families of hymenopterans, with an estimated Australian fauna of around 2000 species within 22 subfamilies and 215 genera. Ichneumons are small to large (2–120 mm in body length) with an elongate, round to thin body, and the first segment of the gaster is divided. Antennae are long and have more than 16 segments. The legs are long and gangly, and the ovipositor is usually prominent, thin and often as long or longer than the body. Ichneumonids can be distinguished from the closely related Braconidae by the forewing venation. In the ichneumons there is an extra vein creating a cell not present in braconids, as shown in Figure 53.

Ichneumon wasp, *Gotra* sp., **Ichneumonidae** (16 mm long)

The vast majority of ichneumons are parasites. They specialise on lepidopteran and hymenopteran (sawfly) larval hosts and those of six other orders. As the ovipositor can be up to

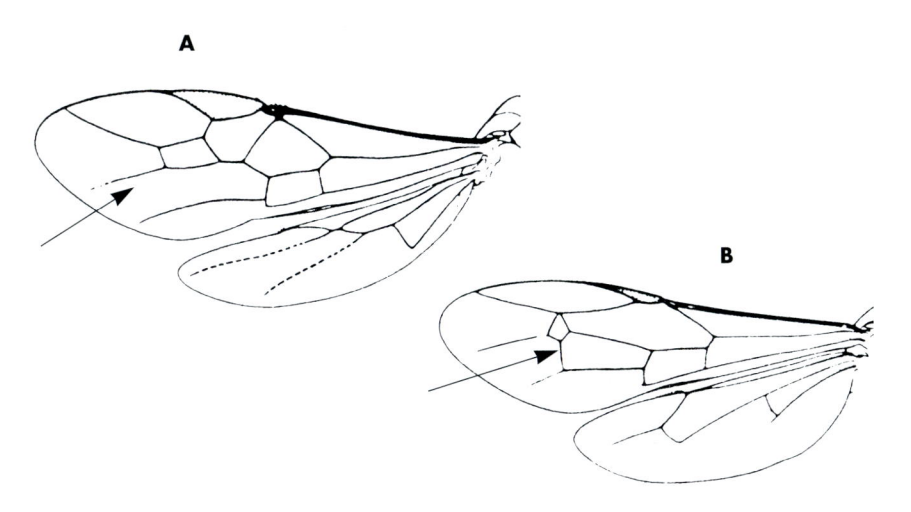

Figure 53. Wing venation in braconid and ichneumon wasps: (a) Braconidae—the arrow indicates the absence of the extra vein; (b) Ichneumonidae—the arrow indicates the presence of the extra vein which creates a cell not found on braconid wings.

Ichneumon wasp,
Xanthopimpla sp.,
Ichneumonidae
(20 mm long)

three times the body length, the females of some species can reach burrowing hosts deep within wood. Once a female finds the host's food plant, she searches the surface with the antennae. When detected, eggs are then laid either on or inside the host's body, and the ichneumon larvae either burrow into the host or feed on it externally. The host usually survives until the parasite is fully grown, at which point the parasite either pupates inside the host or spins a cocoon outside. Adults can be found feeding at flowers, and the majority of species are found away from the arid interior. Ten species of *Gotra* are widespread, and spend their time on trees looking for wood-boring moth larvae to parasitise. In the tropics, this job is often performed by the brightly coloured species of *Xanthopimpla*.

Family Braconidae

This is another large family, whose distinguishing characteristics are discussed above, under the closely

Callibracon sp., **Braconidae** (12 mm long)

WASPS, ANTS, BEES AND SAWFLIES

related Ichneumonidae. Naumann (1991) approximated the size of the Australian fauna at 800 species in 24 subfamilies. They are mostly small to medium-sized, but range from 1 to 80 mm in length. Braconids parasitise many insect orders as internal or external parasites of mainly larvae, but eggs and adults are attacked as well. Lepidoptera, Coleoptera, Orthoptera, Hemiptera (aphids) and other Hymenoptera are the major hosts. Various species of braconids are important biocontrol agents of pest insects. Species of the genus *Callibracon* can be seen inspecting damaged and fallen trees, seeking out longicorn beetle larvae to parasitise. Another tree-trunk frequenting genus is the very colourful *Chaoilta*.

Family Diapriidae

This is the largest family (325 species) within the parasitic superfamily Proctotrupoidea. They are small (1–6 mm in length) and similar

Chaoilta sp., **Braconidae** (20 mm long)

to the many species of the large parasitic superfamily Chalcidoidea in that the wing venation is often very reduced. However, diapriids can be distinguished by having thread-like antennae, unlike the elbowed antennae of chalcidoids (see Figure 54). Diapriids also have a characteristic shelf-like extension of the front of the head which supports the antennae, and the first segment of the gaster is more than half the length of the whole gaster. These wasps frequent humid shady areas where the sometimes wingless females search for late instar larval and pupal hosts of various insect orders. The Australian diapriids largely specialise on parasitising flies.

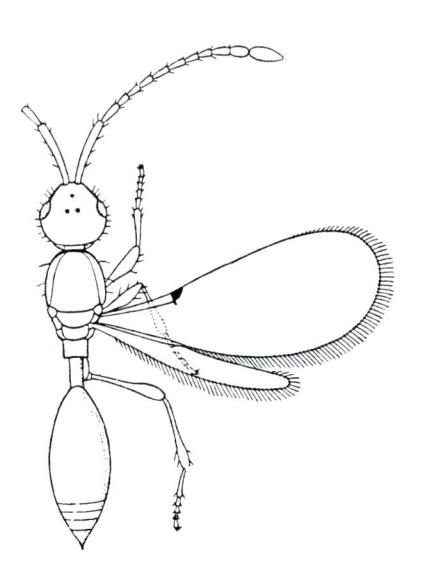

Figure 54. *Trichopria* sp., (family Diapriidae). Note the large first abdominal segment and the antennal insertion point (2 mm). [R.J. Tillyard]

Superfamily Chalcidoidea

This is a very diverse superfamily with 3646 Australian species in 20 families. Most are minute to small (less than 4 mm in body length), though they range from the extremes of 0.15 to 40 mm. They have a range of colours, and many are strongly metallic. The forewings have very reduced venation and the hind wings only one vein (see Figure 56). The pronotum does not extend back to the forewing insertion point as seen from above—this gives the appearance of two body segments forward of the wings, while other similar wasp families show only one segment. The antennae are generally short, prominently elbowed, and 13-segmented. Members of the main family Chalcididae stand out by having very enlarged hind legs.

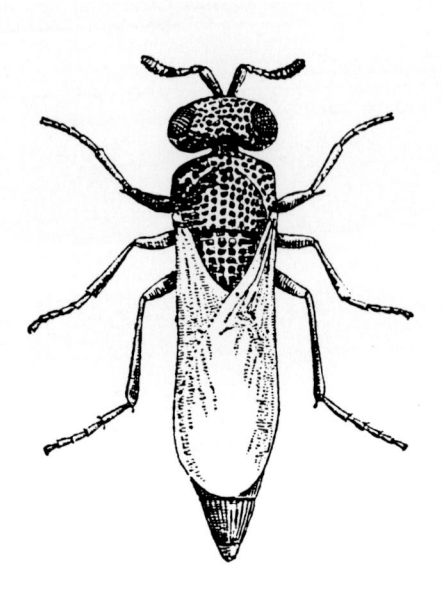

Figure 55. *Scelio* sp. (family Scelionidae) 4 mm. [Dept. Agric. NSW]

Family Scelionidae

This is another important family (445 species) of tiny (0.5–8 mm in body length) parasitic wasps. Most are black and coarsely sculptured; the antennae are elbowed and 11- or 12-segmented; wing venation is very reduced to no venation in the hind wing and rarely more than one vein in the forewing; the gaster, when viewed from above, appears flattened and has a distinct lateral margin. Scelionids differ from chalcidoid wasps by the pronotum extending back to the forewings, and the usually total lack of venation of the hind wing. All are parasites of eggs. The majority of hosts are in the order Orthoptera; but Hemiptera, Coleoptera and Mantodea are also parasitised. The genus *Baeus* has minute, compact, wingless females which parasitise the eggs of spiders. The 32 species of *Scelio* (see Figure 55) are parasites of acridid grasshopper eggs, and are important in controlling the numbers of the plague locust.

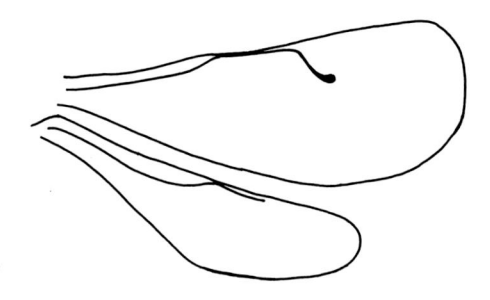

Figure 56. Typical chalcidoid wings, with what appears as only one vein per wing.

The biologies of the chalcidoids are diverse, though most species are parasites of other insects, or parasites of other parasites (hyperparasites). The families Trichogrammatidae (140 species) and Mymaridae (270 species) include some of the smallest insects in the world—down to 0.15 mm in body length. Many species have hair-fringed wings, which give them the name fairy flies, though being so small they

WASPS, ANTS, BEES AND SAWFLIES

Anastatus sp., **Eupelmidae** (3 mm long)

Chalcid wasp, *Brachymeria* sp, **Chalcididae** (6mm long)

Many species are of economic importance as they parasitise insect pests. Species of the genus *Anastatus* in the typical egg parasite family Eupelmidae are among those used in biocontrol.

are moved around as much by wind as by flight. All species of these two families are egg parasites, but other chalcidoid families also parasitise larvae and pupae of Lepidoptera, Coleoptera, Diptera and Hymenoptera. Some families are phytophagous, either making galls, feeding on leaves, or feeding in seeds. Several families, but especially the Agaonidae (50 species), live in the flowers and fruit of figs (*Ficus* sp.) and are essential pollinators of these trees.

(b) Suborder Apocrita — Aculeate Group

Cuckoo Wasps (Family Chrysididae)

This family of 76 species belongs to the first of the aculeate superfamilies, even though the larvae are parasitic and the adults do not sting. They are small to medium-sized (2–22 mm in body length), usually bright green metallic coloured wasps, often seen hovering in the garden or bush. The

Cuckoo wasp, **Chrysididae** (10 mm long)

187

Spider wasp, *Hemipepsis* sp., **Pompilidae** (35 mm long)

body is stocky with the waist only a thin break in the body outline. The abdomen has only a few visible segments, and a concave area on the

underside allows the wasp to roll into a ball for protection. Most chrysidids are parasites of other aculeate nest-building wasps (Sphecidae and Vespidae) which will attack these intruders. Other species are parasites of stick-insect eggs.

Superfamily Vespoidea

This is the largest superfamily of wasps, with close to 5000 Australian species placed in seven families which include the ants. Most are medium to large, ranging from 1 to 40 mm. Colours are usually bright with many black, orange, yellow combinations, and the females have stings. Unlike the bees (Apoidea), no branched or plumose hairs are found on the body. The vespoids also differ from the Apoidea and the Sphecoidea by the

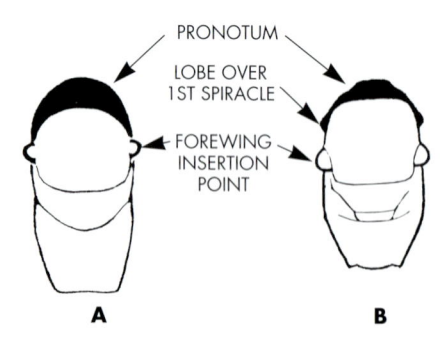

Figure 57. The thorax viewed from above to illustrate the difference in the shape of the pronotum between: (a) the Vespoidea where it meets the forewings; and (b) the Sphecoidea and Apoidea, where it does not. [after Chinery]

WASPS, ANTS, BEES AND SAWFLIES

shape of the pronotum. Viewed from above, it looks semi-circular and extends back to the forewing insertion point. In these other superfamilies it is smaller, looking like a collar which does not extend back to the forewings (see Figure 57). Following is an introduction to six of the seven vespoid families.

Spider Wasps (Family Pompilidae)

This family of 231 species includes mainly medium to large wasps, up to 35 mm long, with strong often black and orange colouring, and tinted wings. Pompilids can be distinguished from other vespoids in the way they hop and flick their wings when hunting, and especially by holding their wings flat over the top of the body at rest, making the wings look larger than in families where they are folded neatly along the sides of the body. The females hunt spiders which they paralyse and store in burrows with one egg per spider. Some species dig their own burrows, but others use other wasps' burrows, or even the spider's own burrow. There are also species which specialise in stealing prey from other pompilid wasps. The genus *Hemipepsis* has one of the largest species at 35 mm long.

Velvet Ants (Family Mutillidae)

These are small to medium-sized parasitic wasps (500 species) sometimes known in many countries as velvet ants because of the wingless females' resemblance to hairy ants. Mutillids are usually hard with coarse punc-tures, and the numerous setae are often arranged in particular colour patterns on the females. The males are usually larger than the females, and have a dark coloured body and tinted wings. The females can be seen scurrying on the ground in exposed habitats looking for host nests. The known Australian species all parasitise social wasps and bees and most species

Female 'velvet ant' *Ephutomorpha* sp., **Mutillidae** (15 mm long)

are placed in the large genus *Ephutomorpha*.

Flower Wasps (Family Tiphiidae)

Like the mutillids, this large family (750 species) of parasitic wasps also contains mainly species with wingless females. Tiphiids are small to large

Female flower wasp, *Hemithynnus* sp., **Tiphiidae** (20 mm long)

(2–40 mm in body length), and usually shiny and brightly coloured. They can often be seen flying together between flowers, with the male carrying the smaller female while mating. The females are stout, with the segments of the thorax pronounced, smooth, and narrower than the abdomen, and the mid and hind legs are spiny and adapted for digging. (In contrast, the mutillid wingless females have a broad sculpted thorax with indistinct segments, a hairy body and simple running legs.) The habits of the family in Australia are not well known, but many species parasitise the larvae of burrowing beetles, and one large species is a parasite of mole crickets.

Hairy Flower Wasps (Family Scoliidae)

Black and orange colours dominate in these large and very hairy wasps (25 species). Males and females are both winged and stoutly built. The females have spiny legs for digging into wood or the soil in search of beetle larvae to parasitise. Both sexes can be seen feeding on flowers, especially eucalyptus blossoms. *Trisciloa ferruginea* is a giant 40-mm wasp, and possesses a formidable set of mandibles.

Paper Wasps, Potter Wasps and Others (Family Vespidae)

These are among the best known Hymenoptera (324 species). The family contains most of the social wasps, whose paper and mud nests can be seen in many gardens. Their strongest identifying characteristic is a notch in the inner margin of the eyes on either side of the antennae. The pronotum reaches the forewings which are folded along the body at rest, and the mandibles of many species cross over. A common place to spot many species is at the margins of water bodies, as the females of mud wasps need a

Hairy flower wasp, *Trisciloa ferruginea*, **Scoliidae** (40 mm long)

WASPS, ANTS, BEES AND SAWFLIES

Potter wasp, *Paralastor conspiciendus*, **Vespidae** (12 mm long)

Potter wasp, *Delta campaniforme*, **Vespidae** (24 mm long)

constant supply of water to construct their nests. There are four subfamilies in Australia with distinct habits.

The 300 species of the mud or potter wasps belong to the subfamily Eumeninae. They are solitary and build nests of mud, or make burrows, or modify the existing holes of other insects. They can be various colours, up to 30 mm long, and in Australia most provide lepidopteran caterpillars as food for their larvae. The genus *Delta* has many large, handsome species which build characteristic jug-shaped mud nests, often in the eaves of houses. With 130 species, the genus

Paper wasps, *Ropalidia* sp., **Vespidae** (12 mm long)

Paper wasps,
Polistes sp.,
Vespidae
(14 mm long)

Paralastor is the most typical form of this subfamily.

The 33 species of the subfamily Masarinae also build mud nests or use burrows, but provision these with nectar and pollen for their larvae.

With 34 species, the subfamily Polistinae contains two genera of native, colonial paper wasps, *Ropalidia* and *Polistes*. Each nest can have from one to many hundreds of individuals, and contains one fertile queen and a population of workers (many of which are suppressed queens). The nest structure varies with species, but usually consists of layers of combs, sometimes spiralling, and sometimes enclosed in an extra layer. Adults feed on nectar, but they hunt caterpillars to feed their larvae, and can give a painful sting in defence of the nest.

The Vespinae is a northern hemisphere subfamily with two introduced species in Australia. *Vespula germanica* is the European wasp which lives in large colonies, either underground or in cavities, and feeds its larvae on a variety of insect prey. It is considered a pest because of its very aggressive behaviour and potent sting.

Ants (Family Formicidae)

The ants are one of the most familiar and important group of insects. In 1985 Taylor and Brown catalogued the Australian fauna and placed them in nine subfamilies, with 95 genera and an estimated total of 4000 species. The principal characteristic of ants is the waist. It is composed of usually one or two knobs (which are the first one or two segments of the abdomen) rather than the single petiole of other hymenopterans. The antennae have a distinctive elbow, with the scape nearly as long as the segments past the elbow, and the mandibles are prominent.

Ants live in colonies made up of several castes and subcastes: winged males, winged females, and several subcastes of workers. Colonies usually have hundreds to several thousand individuals, though the range is from less than 100 to more than a million. Winged males and females appear only at times suitable for mass mating flights. The males die after mating, and the females shed their wings and usually start a new colony, or sometimes join an existing one. To begin a colony the female lays a small number of eggs and then feeds the larvae to maturity.

WASPS, ANTS, BEES AND SAWFLIES

Green tree ants with 'woven' nest, *Oecophylla smaragdina*, **Formicidae** (9 mm long)

These first workers then take over the functions of nest building and look after this first queen whose only remaining function is to lay eggs.

The worker subcastes include the worker minors which perform nest building, nursery care and food foraging; workers with large heads called soldiers which defend the nest; and intermediate forms in different species. Some species have a specialised subcaste called repletes which

Bull ant, *Myrmecia nigrocincta*, **Formicidae** (20 mm long)

act as storage vessels for the colony nectar supply.

Ant nests are usually in the ground, but can be in trees and even among silk-joined leaves in the green tree ant, *Oecophylla smaragdina*, which dominates in the tropics. The best known Australian ants are the aggressive, stinging bull ants, in the large endemic genus *Myrmecia*. Among the ground nesting ants, the meat ants, *Iridomyrmex*, are well-known red and black foragers which dominate the more arid landscapes. One of the largest genera is *Polyrhachis*, which includes metallic species and the 'golden-bum' ants.

Ants have a varied diet, from predators and scavengers, to plant eaters, fungus eaters and other specialisations or combinations of these. Some species of plants, especially in the tropics, have a special relationship with particular ant species. The so-called 'ant-plants' have special chambers colonised by the ants and sometimes even provide food for them in return for protection from herbivores. Another ant relationship involves specialised species in eight other insect orders called myrmecophiles. These live within the nests of ants either as predators or scavengers, or are actively encouraged by the ants as they produce sweet secretions favoured as ant food. Some ants actually 'farm' aphids, scale insects and some butterfly larvae for this purpose.

Mud Daubers and Sand Wasps (Family Sphecidae)

This large family contains many of the medium to large (1.5–40 mm in body length) predatory wasps that are likely to be encountered. Cardale (1985) catalogued 600 Australian species in 54 genera. Sphecids are closely related to bees; though bees differ by having branched body hairs, broad hind tarsi, and using pollen and nectar as a food supply for their larvae. (See Vespoidea on p. 188, for

Meat ants, *Iridomyrmex* sp., **Formicidae** (8 mm long)

Golden-bum ant, *Polyrhachis* sp., **Formicidae** (10 mm long)

differences from other predatory wasp families.)

Sphecid lifestyles range from solitary parasitic to subsocial predatory. They prey on a variety of insect orders and spiders, with some of the seven subfamilies specialising in one order only. Nests are dug into the ground, various mud constructions, natural openings, or the existing burrows of their hosts. Some genera even use the tool of a stone held in their mandibles to help construct the burrow. The adults, which feed on nectar, can be seen either dragging their paralysed prey or flying with it in their mandibles or with it impaled on their sting. Most species make a burrow for each larva, but subsocial species with some division of labour and colonies in the ground occur in several genera such as *Bembix*, which fills its burrow with fly hosts for the larvae.

Bees (Superfamily Apoidea)

This large group has an estimated 1600 species in seven families and, except for a few parasitic species (on other bee nests), makes universal use of pollen and nectar as food for the larvae. Characteristics include branched hairs on the body, and a pollen 'basket' which is usually on an enlarged first tarsal segment and tibia of the hind leg or on the undersurface of the abdomen. Some bees in the family Colletidae carry both nectar and pollen in an internal crop. Bees have the mouthparts modified into a 'tongue' (glossa), and can be roughly grouped into long-tongued and short-tongued families. Most bees are solitary. Each female digs a vertical

Sand wasp, *Bembix* sp., **Sphecidae** (14 mm long)

WASPS, ANTS, BEES AND SAWFLIES

Native bee, **Halictidae** (8 mm long)

ABOVE: Carpenter bee, *Xylocopa aruana*, **Anthophoridae** (22 mm long)
RIGHT: Bluebanded bee, *Amegilla* sp., **Anthophoridae** (10 mm long)

burrow in the ground or in wood or plant stems. Cells containing the nectar and pollen larval food are placed in the burrow or in adjoining corridors. Many bee species nest in small to large groupings with individual cells in a large burrow with a shared entrance. A smaller number are truly social, with males, egg-laying queens and sterile workers in each colony. Colonies can include many thousands of workers and are important pollinators as well as the source of honey and bees wax.

Short-tongued Bees (Family Colletidae)

The largest Australian bee family, with over half the described species, is the Colletidae. These are short-tongued bees, often sparsely haired, and specialising on the nectar and pollen of myrtaceous plants such as *Eucalyptus*. The large subfamily Euryglossinae is endemic with some very small species that can be completely yellow, and even the rare colour white.

Burrowing Bees (Family Halictidae)
With 382 species, the Halictidae is a varied family of solitary and colonial bees which usually nest in burrows in the ground. These small, often dark and sometimes metallic-coloured bees carry pollen both on the back legs and on ventral abdominal hairs.

Family Megachilidae
These bees are long-tongued and include the leaf-cutter bees which line their nests with cut leaves and frass. They carry pollen on the underside of the abdomen.

Carpenter Bees (Family Anthophoridae)
The largest Australian bees are the carpenter bees in the family Anthophoridae. The genus *Xylocopa* has large, very hairy and often metallic species which make nests inside dead branches of the grasstree *Xanthorrhoea*. The banded species of *Amegilla* make their nests underground.

WASPS, ANTS, BEES AND SAWFLIES

Sugarbag bee, *Trigona* sp., **Apidae** (5 mm long)

Family Apidae

The family Apidae includes the introduced commercial honey bee *Apis mellifera*, and the small native honey bees in the genus *Trigona*. These small black social bees are variously called 'sugarbag bees' and 'sweat bees'. They are very common on blossom and do not sting. Many books have been written on the honey bee, and Cardale (1993) has catalogued the described genera and species of Australian bees, including information on the plants they use.

Further Reading

Cardale, J.C. (1993), 'Apoidea' in *Zoological Catalogue of Australia*, vol. 10, *Hymenoptera: Apoidea*, AGPS, Canberra.

Evans, H.E. and Eberhard, M.J.W. (1973), *The Wasps*, David and Charles, Newton Abbot, UK.

Gauld, I. and Bolton, B. (eds) (1988), *The Hymenoptera*, British Museum of National History, Oxford University Press, Oxford.

Greenslade, P.J.M. (1979), *A Guide to the Ants of South Australia*, South Australian Museum, Adelaide.

Naumann, I.D. (1991), 'Hymenoptera' (ch. 42) in *The Insects of Australia*, CSIRO, Melbourne.

Swarming honey bees, *Apis mellifera*, **Apidae** (12 mm long)

GLOSSARY

ABDOMEN. The rear of the three body divisions of an insect.

ANTENNAE. A pair of jointed sensory organs on the head of all insects; also known as 'feelers'.

ANTERIOR. Forward-facing section.

APTEROUS. Wingless.

ARISTA. The thin bristle-like tip of some antennae, mainly in flies (Diptera). See Fig. 5 (H).

BRACHYPTEROUS. With shorter than normal wings.

CALLUS. A swelling on the thorax of some flies, usually near the base of the wings.

CALYPTER. A lobe or flap at the base of the wings of some flies which obscures the HALTERES from above.

CARINA. A raised ridge, often on the wing covers (ELYTRA) of beetles.

CASTE. One of three or four distinct body forms among social insects, including a reproductive queen, workers, soldiers and reproductive males.

CELL. An area enclosed by veins on wings.

CERCI. A pair of jointed appendages at the tip of the abdomen.

CHRYSALIS. A term used for the PUPA of mainly butterflies and some moths.

CLUBBED. With a swollen, bulbous end. A term applied to the shape of antennae. See Fig. 5 (C).

COCOON. A silken case which encloses the pupa of several insect groups including many moths.

COMPLETE LIFE CYCLE. The growth cycle where the young (the larvae) have a different form from the adult, and undergo a pupal stage to become the adult. See ENDOPTERYGOTE (page 17).

COXA (pl. COXAE). The first basal segment of the insect leg. It is sometimes plate-like and fixed in position. See Fig. 6.

CROCHETS. A ring of minute hooks on the ends of the abdominal prolegs of caterpillars of Lepidoptera.

CUNEUS. A triangular shape on the forewings of certain bugs (especially Miridae), separated by a groove. See Fig. 30.

DEFLEXED. Pointing downwards.

DETRITUS. Broken-up and usually decaying organic matter.

DIMORPHISM. Differences in the appearance of individuals of the same species, especially between the sexes.

DISTAL. Furthest from the body, e.g. the tip of the wing is the distal part of the wing.

DORSAL. The upper or top surface.

DORSOVENTRAL. Between the upper and lower surface; therefore, a dorsoventrally flattened insect is flat against the surface it sits on.

ECDYSIS. The process of moulting or casting off the outer 'skin' in insect growth.

ELYTRA. The hardened forewings that protect the membranous hind wings in beetles (Coleoptera) and are not used in flight.

ENDEMIC. Restricted in distribution to a particular region.

ENDOPARASITE. A parasite living within the body of its host.

ENDOPTERYGOTE. The insect groups whose life cycle involves a larva–pupa–adult sequence. The wings develop on the inside of the larval body and not gradually on the outside as in EXOPTERYGOTES.

EXOPTERYGOTE. The insect groups whose life cycle involves gradual development from nymph to adult. The wings develop on the outside in stages.

EXOSKELETON. The tough, jointed outer covering or skeleton of insects and other arthropods. It is made of chitin; and as it does not grow it has to be shed (moulted) periodically to allow the animal to increase in size.

FEMUR (pl. FEMORA). The third of five segments of the insect leg. It is between the TROCHANTER and the TIBIA and is often the largest segment. See Fig. 6.

FILIFORM. Threadlike, a term applied to the shape of antennae. See Fig. 5 (A).

FLAGELLUM. The main part of the antennae, after the SCAPE.

FRASS. Plant fragments made by wood-boring insects, usually mixed with excreta.

FRENULUM. A set of bristles on the hind wings of many moths which link the wings in flight.

GALL. A growth on plants, often bulbous, caused by the irritation of certain insect groups whose young stages develop inside.

GASTER. The main part of the abdomen in wasps and ants, behind the 'waist' or petiole.

GLOSSA. The mouthparts of bees, also called a 'tongue', formed from lobes on the LABIUM.

HALTERES. The club-shaped balancing organs which all flies (Diptera) have in place of the hind wings.

HEMELYTRA (s. HEMELYTRON). The partly hardened and partly membranous protective forewings of true bugs (Heteroptera) which are still used in flight.

HEXAPOD. Six legged.

HONEYDEW. The sweet secretion of aphids and plant hoppers (Hemiptera) which attracts ants.

HYPERPARASITE. A parasite which parasitises another parasite.

IMAGO. The adult stage of an insect.

INCOMPLETE LIFE CYCLE. The growth cycle where the young (nymphs) are similar in appearance to adults and develop gradually without a pupa stage. See EXOPTERYGOTE.

INQUILINE. An animal living in the nest of another species. In insects, most commonly in the nests of termites and ants.

INSTAR. One stage of growth between moults, from egg to adult.

LABIAL PALPS. One of two sets of jointed palps (palpi) on the sides of the 'mouth' of insects, in front of the MAXILLARY PALPS. See Fig. 3.

LABRUM. A plate or flap at the front of the head also known as the 'upper lip' because it often partly covers the mouthparts. See Fig. 3.

LAMELLATE. Pertaining to antennae with fanlike segments at the tip. See Fig. 5 (E).

LARVA. The immature (often grub-like) stage of ENDOPTERYGOTE insects. See pages 17–18.

LATERAL. Pertaining to the sides.

MANDIBLES. The upper, chewing pair of mouthparts (jaws) of insects, sometimes modified into other shapes.

MANDIBULATE. With biting and chewing mouthparts.

MAXILLAE. The second, partly internal, pair of mouthparts of insects.

MAXILLARY PALPS. One of two sets of jointed palps on the sides of the 'mouth' of insects, behind the LABIAL PALPS. See Fig. 3.

MEMBRANOUS. Pertaining to wings, the usually transparent flying wings, like in dragonflies.

MESOTHORAX. The second (middle) of the three segments of the THORAX.

METATHORAX. The third (rear) of the three segments of the THORAX.

MONILIFORM. Beadlike, a term applied to the shape of antennae. See Fig. 5 (B).

MOULT. To shed the outer 'skin' or exoskeleton in the process of growth.

NYMPH. The immature stage of EXOPTERYGOTE insects. Nymphs usually resemble the adult except for

Tiger moth, *Euchromia creusa*, **Arctiidae** (45 mm across)

lacking functional wings which develop gradually on the outside.

OCELLUS (pl. OCELLI). A simple, single-lensed eye present, often in a pattern of three, in many insect groups.

OVIPOSITOR. The egg-laying apparatus of female insects, concealed in some orders, and very long and conspicuous in others, e.g. wasps and crickets.

PALPS/PALPI (s. PALP/PALPUS). Segmented appendages around the mouthparts which are organs of taste and help manipulate food. See Fig. 3.

PARTHENOGENESIS. Reproduction from unfertilised eggs.

PEDICEL. The two-segmented 'waist' of ants.

PETIOLE. The narrow 'waist' segment of wasps.

PHYTOPHAGOUS. Feeding in or on plants.

PLASTRON. A layer of fine hair on the underside of some aquatic insects, which holds a layer of air.

PROBOSCIS. Extended mouthparts which are often modified into a tube for sucking. The term is most often used for the coiled tubular mouthparts of moths and butterflies.

PROLEGS. Short appendages on the abdomen of some caterpillars that act as legs. Normal legs are attached to the THORAX.

PRONOTUM. The upper (DORSAL) surface of the first segment of the THORAX.

PROTHORAX. The first of three segments of the THORAX.

PROXIMAL. The part of an appendage nearest to the body.

PTERYGOTE. Winged—belonging to the insect orders which are winged (though not all species necessarily have wings). See ch. 1.

PTILINUM. An inflatable structure on the heads of some flies, used to break out of the pupal skin.

PUBESCENT. Covered in fine short hair; downy.

PUPA (pl. PUPAE). The usually inactive stage between larva and adult, found in insects with a COMPLETE LIFE CYCLE. See page 17.

RAPTORIAL. Adapted for seizing and holding prey.

ROSTRUM. The beak-like piercing and sucking mouthparts of all bugs (Hemiptera). See Fig. 4. Also applied to the 'snout' of weevils.

SAPROPHAGOUS. Feeding on decaying organic matter.

SCAPE. The first (basal) and often largest segment of the antenna.

SCLERITE. Any one of the separate plates making up the SEGMENTS of the EXOSKELETON.

SCLEROTISED. Hardened (or armoured) like the ELYTRA of a beetle, compared to the soft tissue of, for example, caterpillars.

SCUTELLUM. The third division of the THORAX as seen from above. It is the very large triangle in some bugs (Hemiptera) between their crossed-over wing covers.

SEGMENT. A division of the EXOSKELETON between flexible joints in the body and appendages.

SERRATE. Sawlike or toothlike, a term applied to the shape of antennae. See Fig. 5.

SETA (pl. SETAE). A stiff bristle-like hair.

STERNITE. Any SCLERITE on the VENTRAL (under the body) surface of an insect.

STERNUM. The VENTRAL or underneath part of any body SEGMENT.

STRIDULATION. Production of sound by rubbing two parts of the insect body, usually a toothed file on legs or wings.

SUTURE. The groove which divides SEGMENTS and SCLERITES on the body.

TARSUS (pl. TARSI). The insect 'foot', the last of five divisions of the leg, and itself divided into one to five segments. See Fig. 6.

TEGMEN (pl. TEGMINA). The toughened, leathery forewing of several orders, especially Orthoptera, Blattodea and Mantodea.

TERGITE. Any SCLERITE found on an upper (DORSAL) surface of an insect.

TERGUM. The upper (DORSAL) part of any body SEGMENT.

THORAX. The middle of the three major body divisions of an insect between head and ABDOMEN. The legs and wings are appendages of the thorax. See Fig. 2.

TIBIA (pl. TIBIAE). The fourth of five segments of an insect leg. It is between the FEMUR and the TARSUS, usually slender and the longest segment. See Fig. 6.

TROCHANTER. The second of five segments of an insect leg. It is between the COXA and FEMUR, usually the smallest and often overlooked segment. See Fig. 6.

TYMPANAL ORGAN. A stretched drum-like membrane which forms the ears of insects like grasshoppers and others.

VENTRAL. On the underside of the body.

VESTIGIAL. Only partly developed, small and non-functional.

VIVIPAROUS. Bearing live young.

MULTIMEDIA BIBLILOGRAPHY

Web Sites

Australian Ants Online
http://www.ento.csiro.au/science/ants/default.htm
Information on Australian Ants

Australian Insect Common Names
http://www.ento.csiro.au/aicn
Search by common or scientific names for the official common names of all Australian insects.

Australian Native Bees
http://www.zeta.org.au/~anbrc/
Good information on stingless and other native bees.

Coleoptera Org
http://www.coleoptera.org
Based in Sydney, this site has links to everything you would want to know about beetles.

The Fly in Your Eye
http://www.viacorp.com/flybook/fulltext.html
All about the Australian Bush Fly.

A Guide to Stick Insects of Australia
http://www.acay.com.au/~pmiller
Find here keys pictures and information on Australian stick insects.

Lepidoptera Larvae of Australia
http://www-staff.mcs.uts.edu.au/~don/larvae/larvae.html
Pictures of many adults and caterpillars of Australian moths and butterflies and outlines of their natural history.

University of Florida Book of Insect Records
http://ufbir.ifas.ufl.edu/
For the world's biggest, fastest or loudest insects.

Books and CD-Roms

Braby, M.F. (2000), *Butterflies of Australia. Their Identification, Biology and Distribution*, 2 volumes, CSIRO Publishing, Melbourne.

Brunet, Bert (2000), *Insects: A Natural History*, Reed New Holland, Sydney.

CSIRO (1997), *Insects—Little Creatures in a Big World*. CD-Rom multimedia, CSIRO Publishing, Melbourne.

Gullan, P.J. and Cranston, P.S. (2000, 2nd ed) *The Insects : An Outline of Entomology*. Blackwell Publishers, London.

Hangay, G. and German, P. (2000), *Nature Guide to Insects of Australia*, Reed New Holland, Sydney.

Harvey, M.S. and Yen, A.J. (1989), *Worms to Wasps*, OUP, Melbourne.

Moulds, M.S. (1990), *Australian Cicadas*, The Australian Museum, Sydney.

Mound, L.A. and Kibby. G. (1998), *Thysanoptera: An Identification Guide*, CABI Publishing.

New, T.R. (1996), *Name that Insect*, OUP, Melbourne.

Shattuck, S.O. (2000), *Australian Ants: Their Biology and Identification*, CSIRO Publishing, Melbourne.

Silsby, Jill (2001), *Dragonflies of the World*, CSIRO Publishing, Melbourne.

Zborowski, P. (2002) *Green Guide to Insects of Australia*, Reed New Holland, Sydney.

INDEX

Page numbers in *italics* refer to photographs and drawings.

abdomen 11, 15, *15*
Abricta 87, *87*
Acerentomidae *40*
Achias australis 144, *144*
Acrididae *19*, 69, *70-1*
Actinus macleayi 114, *114*
Adephaga 110
Adrama 145
Adrisa 93
Aedes aegypti 138, *138*
Aenetus ligniveren 157, *158*
Aeolochroma turneri 166
Aeolothripidae 99
Aetholix flavibasalis 164
Agaonidae 187
Agapophytus 20
Agaristinae 173
Agrotis infusa 173
alderflies 99
Alloeopage cinerea 167
Allomachilis 43
  froggatti 43
Alydidae *22*, 92
Amata marella 173, *174*
Amblycera 80
ambrosia beetles 127
Ambylux wildei 171, *171*
Amegilla 196, 197
Amenia imperialis 147, 148
Amerila rubripes 173, *173*
Amitermes 58
  laurensis 59
  meridionalis 59
Amorphoscelidae *61*
Amphipterygidae 48, *48*
Anastatus 187, *187*
Anisocentropus banghasi 153, *153*
Anisolabididae 63
Anisoptera *47*, 48
Anisozyga pieroides 167
Anopheles 138
Anoplognathus 119

Anoplura 81
ant butterflies 178
ant crickets 66
Anthela 170, *170*
Anthelidae 170, *170*
Anthophoridae *196*, 197
antler flies 143, *144*
antlions 7, 101, *102*, 104, *104*
ants 179, 192
bull *5*, *193*, 194
green tree *193*, 194
meat 194, *194*
golden-bum 194, *195*
Aphididae 84, *85*
aphids 82, 84, *85*
Apidae 198, *198*
Apiomorpha 85, *85*
Apioninae 127
Apiorhynchus 124
Apis mellifera 198, *198*
Apocrita 182
Apoidea 195
Apteropanorpa tasmanica 131
Apteropanorpidae 131
Arachnida 10, *10*
Archaeognatha 42
Archichauliodes 100, *101*
Archimantis 59, 60
Archostemata 110
Arctiidae *22*, 173, *173-4*, *201*
aristate (antennae) 136
armoured scales 86
Arthropoda 10
Ascalaphidae 105, *105*
Asilidae *3*, 140, *141*
Asilinae *141*, 142
assassin bugs 91, *91*
Atalophlebia 44, 45, *45*
Atractomorpha 69, *70*
Auchenorrhyncha 82, 83, 84, 87
auger beetles 121, *121*
Aulacaspis tubercularis 86, *86*
Australembiidae 78
Australembia 77
Austrogomphus prasinus 49, *49*
Austronevra 144, *145*

Austrosialis ignicollis 101

backswimmers 95, *96*
Baetidae 45, *45*
Balta 53, *53*
Batocera wallacei 125
Bactrocera musae 145
Bactrocera tryoni 144
bed bugs 92, *92*
bee flies *141*, 142, *142*
bees 179, 195, *196*
beetles 108
Belenois java 175
Belidae 127, *127*
Belostomatidae *96*, 97
Bembix 195, *195*
Bibionidae 139, *139*
birdwings 175, *176*
Biroella 69, *69*
Bittacidae 131, *132*
Blaberidae 54, *54*
Blattella germanica 54
Blattellidae 53, *53*, 54
Blattidae 53, *53*
Blattodea 52
Blepharotes 140
blowflies *147*, 148, *148*
bluebanded bees *196*
bluebottles 148
blues (butterflies) 178, *178*
bogong moths 173
Bombyliidae *141*, 142, *142*
Bombylius 141
booklice 78, *79*
Bostrichidae 121, *121*
Bostrychopsis jesuita 121, *121*
Brachycera 136
Brachymeria 187
Brachystomella 40
Brachystomellidae *40*
Braconidae 21, *183*, 184, *184*, 185
Brentidae 127, *128*
Brentinae 128
bristletails 42, *43*
brown lacewings *102*, 106, 107
browns (butterflies) 175, *177*

Bruchinae 125
Brunotarsessus fulvus 88, *88*
budworms *172*, 173
bugs *12*, 82
bull ants *5*, *193*
Buprestidae 122, *122*
burnets 162, *163*
burrowing bugs 93, *93*
bush flies 147
butterflies 155, 174

cabbage whites 175
Cacostomus squamosus 115, *115*
caddisflies 151, *152*, 154
Caedicia 66, *67*
Caelifera 68
Caenidae 46
Calamoceratidae 153, *153*
Callibracon 184, 185
Calliphoridae *136*, 147, 148, *148*
Calloodes rayneri 118, 119
calyptrate (flies) 147
camel crickets 68
camouflage 19
Campion 103
Campodeidae 41
Candalides erinus 178, *178*
Cantharidae 120, *120*
Carabidae *31*, 110, *111*
carpenter bees *196*, 197
case moths *158*, 159
caste 56, 180, 192, 195
Castiarina carinata 122, *122*
caterpillars 155
Cephalodesmius armiger 116
Cerambycidae 123, *124*
Cerambycinae 125
Ceratopogonidae 139
cerci 15
Ceriagrion aeruginosum 48, *48*
Ceroplastes rubens 86, 87
Cetoniinae 119
chafers *117*, 119
Chalcididae 186, *187*
Chalcophorinae 122

204

INDEX

Chaoilta 185, *185*
Chathamiidae *152*
Chauliognathus lugubris 120, *120*
Chelisoches 63
Chelisochidae 63, *63*
Chironomidae 139
Choristidae 132
Chortoicetes terminifera 69, *70*
Christmas beetles *118*, 119
chrysalis 17, 155
Chrysididae 187, *187*
Chrysomelidae *124*, 125, *125*
Chrysomelinae 125
Chrysomya bezziana 148
Chrysopidae *102*, 105, *105*, *106*
cicadas 87, *87*
Cicadellidae *19*, 87, *87*, 88
Cicadidae 87, *87*
Cicindela semicincta 111, *111*
Cicindelinae *31*, 110
Cimex lectularius 92, *92*
Cimicidae 92, *92*
class 8
Cleridae 121, *121*
click beetles 119, *119*
Cloeon 45, *45*
clothes moths 158
clubbed (antennae) 13
Coccidae 86, *86*
Coccinellidae 123, *123*
Coccoidea 85
cockroaches *11*, 52, *53*
Coenagrionidae *47*
Coleoptera *18*, *26*, 108
Collembola 39, *40*
Comana mittogramma 163
Comostola chloragyra 167
Complete life cycle 17, 30
compound eyes 12
Coniopterygidae 103
Conocephalus semivittatus 67
cooloola monster 68, 69
Cooloola propator 69
Cooloolidae 68, 69
Coptotermes 57
Corduliidae *47*
Coreidae 91, *91*
Corixidae 96, *96*
Corydalidae *100*, 101
Coryphistes 72
Coscinocera hercules 169, 170
Cosmopterigidae 160, *160*

Cosmozosteria 53, *53*
Cossidae 160, *161*
Cotton harlequin bug 28, 93, *93*
cottony cushion scale 86, *86*
coxa 13
crab lice 81, *81*
crane flies 137, *137*, *138*
Cranopygia 63, *63*
crickets 64
 black field 66
 camel 68
 king 67, *68*
 mole 66, *66*
crusader bugs 91, *91*
Crustacea *10*, 11
crypsis 19
Cryptolaemus 123
Cryptotermes 57
Ctenocephalides canis 134, *134*
Ctenolepisma 43, *43*
cuckoo wasps 187, *187*
Culex fatigans 138
Culicidae *136*, 138, *138*
cuneus 90, *90*
cup moths 161, *162*, *163*
Curculionidae *109*, 125, *126*
Curis 122
Cyaria 22
Cydistomyia 140
Cydnidae 93, *93*
Cylindrachetidae 72
Cyphogastra farinosa 122, *122*

damselflies 46
Danaus plexippus 178
darkling beetles 123
Dasybasis 140, *140*
Delta campaniforme 191
Derbidae 89, *90*
Dermaptera 62
Diapriidae 185, *185*
Diaspididae 86, *86*
Dinotoperla 51, *51*
Diphlebia euphaeoides 48, *48*
Diplacodes haematodes 49, *49*
Diplura 39
Diptera *26*, 135, *136*
diving beetles 111, *112*
dobsonflies 99, *100*
Dolichopodidae 142, *142*
dragonflies 46
drone flies 143, *143*
Drosophila 145, *146*
Drosophilidae 145, *146*

dung beetles 116, *116*
dusty wings 103
Dynastinae 119
Dysphania fenestrata 168
Dytiscidae *109*, 111, *112*

earwigs 62, *63*
Ecnomidae 153
Ectochemus decimmaculatus 128, *128*
Elateridae 119, *119*
elbowed (antennae) 13
Elipsocidae 79
Ellipsidion australe 53, 54
elytra 15
embiids 77
Embioptera 77
emperor gum moths 168, *170*
Endacusta 66, *66*
endopterygote 17, *18*
Endoxyla mackeri 160, *161*
Ensifera 65
Entognatha 39
Eoxenos 129
Ephemeroptera 44
Ephutomorpha 189, *189*
Eriococcidae 85, *85*
Eristalinus punctulatus 143, *143*
Euchromia creusa 201
Eumastacidae 69, *69*
Eumeninae 191
Eupelmidae *187*
Eurema hecabe 175, *177*
European wasps 192
Eurybrachyidae 89, *90*
Eurycnema goliath 74, *75*
Eurymelidae 88, 89
Eurymeloides pulchra 88, 89
Eustheniidae 50
Eutryonia monstrifer 88, *88*
Evaniidae 182, *182*
Exoprosopa 142
exopterygote 17, *17*
exoskeleton 10
Extatosoma tiaratum 75, *75*
eyes 12

fairy 'flies' 186
families 8
femur 13
field crickets 65, 66
fig wasps, 187
filiform (antennae) 13
fireflies 120, *121*

fishkiller bugs 96, 97
Flatidae 89, *89*
flea beetles 125
fleas 133, *134*
flesh flies 148, *148*
flies *12*, 135, *136*
flower wasps 189, *189*
Forficula auricularia 63, *63*
Forficulidae 63, *63*
Formicidae *5*, *191*, 192, *193*, *195*
fritillaries 175
fruit flies 144, *145*
Fulgoridae *19*, 89, *89*

Galerucinae 125
Gelechiidae 160
Gelechioidea 159
Geometridae 165, *166*, *167*, *168*
Gerridae 94, *94*
Gerromorpha 94
giant thrips 98
ghost moths 157
Glycaspis 84, *85*
Glyphodes canthusalis 165
goat moths 160
Gomphidae 49, *49*
Goniaea 71, *72*
Gonocephalum 123, *123*
Gotra 183, 184
Gracillariidae 159
Graphium 175
grasshoppers *19*, *38*, 64, 67, 71
grass yellows 175, *177*
green lacewings *102*, 105, *105*, *106*
green tree ants *193*, 194
Gripopterygidae 51, *51*
ground beetles 110, 111
Gryllacrididae 67, 68
Gryllacridoidea 67
Gryllidae 65, 66
Grylloidea 65
Gryllotalpa 66, *66*
Gryllotalpidae 66, 66
gumtree hopper 88
Gynoplistia 137, *137*
Gyrinidae 112, *113*
Gyromantis 61, 62

Haematopinus 81
Haematopinidae 81
hairy flower wasps 190, *190*
Halictidae *196*, 197
hanging flies 131
Hapsidotermes maideni 57, *58*
Harpobittacus tillyardi 131, *132*

205

A FIELD GUIDE TO INSECTS

hatchet wasps 182, *182*
hawk moths 170, *171*
Helicopsyche 152
heads 11, 12, *12*
Helicopsychidae *152*, 154
Helicoverpa armigera 172, 173
Heliothrips 98
Hemerobiidae 106, *107*
Hemipepsis 188, 189
Hemiptera 17, 26, 82, 84
Hemithynnus 189
Heoclisis 104, 105
Hepialidae 156, *157*
hercules moths 169, 170
Hesperiidae 174, *174*
Heterojapygidae 41, *41*
Heterojapyx 41, *41*
Heteroptera 82, 90
Heteropternis 71, 72
Hippotion scrofa 170, *171*
Hispellinus 125
Hispinae 125
Histeridae 113, *114*
Hololepta 114, *114*
honey bees 198, *198*
hoppers 82
horse flies 139
house flies 147
hover flies 143, *143*
Hydaticus parallelus 112, *112*
Hydrobiosidae 153
Hydrometra 94, *95*
Hydrometridae 94, *95*
Hydrophilidae 112, *113*
Hydrophilus pedipalpus 113
Hydroptilidae 153
Hymenoptera 179
Hypochrysops 178

Icerya purchasi 86, *86*
ichneumon wasps 183, *183*, *184*
Ichneumonidae 183, *183*, *184*
Idolothrips marginata 99
Illeis galbula 123, *123*
imago 16
incomplete life cycle 17, 30
inquilini 109, 194
Insecta 8, 10
instars 16
Iridomyrmex 194, *194*
Ischnocera 80
Isoptera 55
Italochrysa insignis 105
Ithonidae 103

Jappa 45, *45*
Japygidae 41

jewel beetles 122, *122*
jewel bugs 93
jezabels 175
jumping spider *23*

Kalotermitidae 57, *57*
Katianna 40
katydids 22, 66, *67*
king crickets 67, *68*

labial palps *11*, 12
Labidura truncata 63
Labiduridae 63
labrum 12
lacewings 101, *102*, *105*, *106*
Lactura erythractis 162, *163*
ladybirds 123, *123*
lamellate (antennae) 13, 116
Lamiinae 125
Lampyridae 120, *121*
lantern flies 89, *89*
Laphria 140, *141*
Largidae 91, *91*
larva 17
Laxta 54
leaf beetles *124*, 125, *125*
leafhoppers 82, 87, *87*, 88
leaf insects 73, *75*
leafroller moths 160, *161*
Ledromorpha planirostris 19, 87, 88
legs 13, *14*
Lepidoptera 155
Lepismatidae 43, *43*
Leptoceridae *152*, 153, *154*
Leptophlebiidae *44*, 45, *45*
Leptopius quadridens 126, *126*
Lestoidea conjuncta 48, *48*
Lestoideidae 48, *48*
Lethocerus 96
Libellulidae *47*, 49, *49*
lice 80, *81*
life cycle 17, *17*, *18*, 30
Limacodidae 161, *162*, *163*
Limnometra cursitans 94, *94*
Liphyra brassolis 178
Liposcelidae 79
Liposcelis 79, *79*
locusts 69, *70*
longicorn beetles 123, *124*
longlegged flies *142*

loopers 166, *167*, 168, *168*
Lucanidae 115, *115*
Lucilia cuprina 148, *148*
Luciola 121, *121*
Lycaenidae 178, *178*
Lycidae *21*, 119, *120*
Lygaeidae 90, *91*, *92*
Lymantria antennata 174, *174*
Lymantriidae 174, *174*
Lyramorpha parens 94, *94*

Macrobathra 160, *160*
Macrogyrus elongatus 112, *113*
Macropanesthia rhinoceros 54, *54*
Macrosiphum rosae 85, *85*
Malleecola myrmecophila 115
mandibles 12
mango scale 86, *86*
Mantidae 59, 60, *60*, *61*
mantis flies 101, 103, *103*
Mantispidae 103, *103*
Mantodea 59
march flies 139, *139*, 140
Margarodidae 86, *86*
Mastachilus 116, *116*
Mastotermes darwiniensis 57
Mastotermitidae 57
maxilla 12
maxillary palps 12
mayflies *15*, 44, *44*
mealworm beetles 123
mealybugs 84, 85
meat ants 194, *194*
Mecoptera 131
Mecynognathus damelii 110, *111*
Mecynothrips 98
Megacephala 31
Megachilidae 197
Megacrania batesii 75, *76*
Megaloptera 99
Megapodagrionidae *47*
Meinertellidae 42
Melanitis leda 177, *178*
Melolonthinae 119
Membracidae 88, *88*
Mengenillidae *129*, 130
Mengenillidia 130
Meropeidae 132
Merothripidae 99
mesothorax 13
metathorax 13

Metriorrhynchus 21, 120, *120*
Micromus 102 tasmaniae 107, *107*
Microvelia 95, *95*
Micropezidae 146, *146*
Mictis profana 91, 92
midges 139
millipedes 10, *10*
Mimegralla 146
mimicry 20
Miridae *21*, 90, *90*
mites 10
Moerarchis australasiella 158, 159
mole cricket 66, *66*
Molytria 54, *54*
moniliform (antennae) 13
Monistria pustulifera 69, *70*
monkey grasshoppers 69, *69*
Monophlebulus pilosior 86, *86*
Morabinae 69
Morphotica mirifica 160, *160*
mosquitoes 138, *138*
moth lacewings 103
mud dauber wasps 194
Musca domestica 147
Muscidae *137*, 146, 147
Mutillidae 189, *189*
Myiodactylus 107, *107*
Mymaridae 186
Myriapoda 10, *10*
Myrmecia 5, 194 *nigrocincta* 193
Myrmecolacidae 130
Myrmecophilidae 66
Myrmeleontidae *7*, *102*, 104, *104*
Myxophaga 110

Nasutitermes triodiae 58, 59
native bees *196*
Neelipleona 40
Nematocera 136
Nemopteridae 104
Neomantis australis 61, *61*
Neomyia 146, 148
Neotermes 57
Nephrotoma australasiae 137, *138*
Nepidae *96*, 97
Nepomorpha 95
Nesoxypilus albomaculatus 61, 62
Neuroptera 101, *102*
Nicoletiidae 43

INDEX

Nocticolidae 55
Noctuidae 171, *172, 173*
Notoligotomidae 78
Notonectidae 95, *96*
Notonemouridae 51
Nunkeria 67
  brachis 68
nymph 17
Nymphalidae 175, *175, 177*
Nymphidae 107, *107*

ocelli 12
Ocybadistes 174, *174*
Odonata 46
Odontoceridae *152*
Oecetis 153
Oecophoridae *21*, 159, *159, 162*
Oecophylla smaragdina 193, 194
Oligotoma 77, 78
Oligotomidae 77, 78
Ophiorrhobda phaesigma 161, *161*
Opodiphthera eucalypti 170, *170*
order 8, 32
Orthetrum caledonicum 49
Orthodera ministralis 61, *61*
Orthoptera *38*, 64
Osmylidae 106
owlflies 105, *105*

Palorus 123
Panesthia 54
paper wasps 190, *191, 192*
Papilio
  aegeus 175
  ulysses 175
Papilionidae 175, *175, 176*
Paracalais 119, *119*
Parajapygidae 41
Paralastor conspiciendus 191, 192
Paraoxypilus 61, 62
Parasipyloidea *19*, 76, *76*
Pareremus 67, 68
Paropsis obsoleta 125, *125*
Passalidae 115, *116*
Pediculidae 81
Pediculus humanus 81
Penalva 68
Pentatomidae *17*, 91, 92, *92*
Pentatomoidea 92
Penthea pardalis 124, *125*

Perga 181, 182
Pergidae *181*, 182
Petalura ingentissima 49
Petaluridae 49
Periopta ardescens 173, *173*
Periplaneta
  americana 53
  australasiae 53, *53*
Phalidura 126
Phasmatidae *19, 74, 75, 75, 76*
Phasmatodea 73
Philanisus plebeius 152
Philopteridae 80
Phlaeothripidae 98, *98*
Phlogistus 121, 122
Phthiraptera 80
Phylliidae 75
Phylliinae 75
Phyllium 75
Phyllotocus apicalis 118, 119
phylum 10
Physopelta australis 91, *91*
Pieridae 175, *177*
Pieris rapae 175
Pingasa chlora 166
pinhole borers *126*, 127
pink wax scale 86
Pinophilus 114, *114*
planthoppers 89, *90*
Platybrachys 89, *90*
Platypodinae *126*, 127
Platypsectra analis 181, 182
Platystomatidae 143, *144*
Plecia ornaticornis 139, *139*
Plecoptera 50
plume moths 164, *164*
plumose (antennae) 13
podsucking bugs 92, *92*
Poecilometis 91, *93*
Polistes 192, *192*
Polistinae 192
Polyphaga 110
Polyphagidae 55
Polyrhachis 194, *195*
Polysastra costatipennis 124, 125
Polystigma punctata 119
Polyzosteridae 53
Pompilidae *21, 188*, 189
potter wasps 190, *191*
powderpost beetles 121
praying mantis 59
Prioninae 125
Pristhesancus 91, *91*

Problepsis apollinaria 166
proboscis 12
Projapygidae 41
prolegs 13
pronotum 13
prothorax 13
Protura 39
Pselaphidae 114, *115*
Pseudaegeria 159, 160
Psocidae 79
psocids 78, *79*
Psocomorpha 79
Psocoptera 78
Pseudaegeria 21
Psychidae *158*, 159
Psychopsidae 107
Psyllidae 84, *85*
Pterophoridae 164, *164*
Pthiridae 81
Pthirus pubis 81, *81*
Pulex irritans 133
Pulicidae 134, *134*
pupa 17
Pygidicranidae *63*
Pygiopsyllidae 134
pygmy grasshoppers 72, *72*
Pygospila 165
Pyralidae 164, *164*, 165
Pyrgomorphidae *38*, 69, *70*
Ranatra 96
Raniliella 71, *72*
Reduviidae 91, *91*
Rentinus dilatatus 89, *89*
Rhaphidophoridae 68
Rhinotermitidae 57
Rhinotia hemistictus 127, *127*
Rhynchaenus 126
Rhyzobius 123
Riptortus 22
  serripes 92
robber flies *3*, 140, *141*
Rodolia 123
Ropalidia 191, 192
rose aphids 85, *85*
rostrum 12, 125
rove beetles 114, *114*
Rutelinae 119
Rutilia 149, *149*

Sagra 125
Sagrinae 125
Salticidae 23
sand flies 139
sandgropers 72
sand wasps 194, *195*
Sarcophagidae 148, *148*
Saturniidae 168, *169, 170*
sawflies 179, 180, *180, 181*

scale insects 82, 84
  cottony cushion 86, *86*
  mango 86, *86*
  pink wax 86, *87*
scape 13
Scarabaeidae *18, 109*, 116, *116, 117, 118*
Scarabaeinae 116
Scelio 186, *186*
Scelionidae 186, *186*
Sciapus 142, 143
Scoliidae 190, *190*
Scolytinae 127
scorpion 10
scorpion flies 131, *132*
Scutelleridae *28*, 93, *93*
scutellum 92
screw-worm flies 148
seed bugs 90, *91*
seed weevils 125
Selenothrips 98
Senostoma 150, *150*
serrate (antennae) 13
shield bugs 92, *92*
Sialidae 101
Sigmatoneura formosa 79
silky lacewings 107
silverfish 42, *43*
Sigmatoneura formosa 79
Siphanta 89, *89*
Siphonaptera 133
Sirex noctilio 182
Siricidae 182
skippers 174, *174*
Sminthuridae *40*
soldier beetles 120, *120*
Speiredonia zamis 172, 173
Sphecidae 194, *195*
Sphingidae 170, *171*
spiders 10, *10*
spider wasps *188*, 189
spitfire grubs *181*, 182
springtails 39, *40*
stag beetles 115, *115*
Stangeia 164
Staphylinidae 114, *114*
Stenopelmatidae 67, *68*
Stenopsychidae 153, *153*
Stenopsychodes 153, *153*
Sternorrhyncha 83, 84
stick insects 73
Stigmodera 122
stilt-legged flies 146, *146*
stink bugs 93
Stomoxys calcitrans 148
stoneflies 50

207

Stratiomyidae 20, *136*
Strepsiptera 129
Stylopida 130
Stylopidae 130
stylopids 129, *129*
sugarbag bees 198, *198*
Suphalomitus 105, *105*
swallowtails 175
swift moths 156
Symphypleona 40
Symphyta 180, *180*
Syndipnomyia 20
Syrphidae 143, *143*

Tabanidae 139, *139*, *140*
Tachinidae 149, *149*, *150*
tarsal claws 14
tarsi 14
Tectocoris diophthalmus 28, 93, *93*
tegmen 15
Teleogryllus 66
Tenebrio 123
Tenebrionidae 123, *123*
Tenodera 60, *60*
Tephritidae 144, *144*
termites 55
  giant northern 57
  magnetic 58
  spinifex 58
Termitidae 56, 57, *58*

Termopsidae 57
Tessaratomidae 94, *94*
Tetrigidae 72, *72*
Tettigoniidae 22, 67, *67*
Tettigonioidea 67
Therevidae 20
thorax 11, 13
threadwings 104
Thripidae 98, *98*
thrips 97
Thrips imaginis 98, *98*
Thysanoptera 97
Thysanura 42
tibia 13
tibial spurs 14
ticks 10, *10*
tiger beetles *31*, 110, *111*
tiger moths 173, *201*
Tineidae 158, *158*, *162*
Tineola bisselliella 158
Tiphiidae 189, *189*
Tipulidae *136*, 137, *137*, *138*
Titanolabis colossea 63
Torbia 66, *67*
Tortricidae 160, *161*, *162*
treehoppers 19, 87, *88*
Tribolium 123
Trichaulax macleayi *118*, 119

Trichodectidae 80
Trichogrammatidae 186
Trichopria 185
Trichoptera 151
Tridactylidae 72, *72*
Tridactylus australicus 72, *72*
Trigona 198, *198*
Triplectides 153, *154*
Trisciloa ferruginea 190, *190*
Troctomorpha 79
Trogiomorpha 79
Trogloblattella nullarborensis 52
Troides priamus 175, *176*
Tumulitermes hastilis 59
tussock moths *174*

Urnisiella 71, 72

Valanga irregularis 69, *70*
Veliidae 94, *95*
velvet 'ants' 189, *189*
Vespidae *180*, 190, 191, *192*
Vespoidea 188
Vespula germanica 192
vinegar flies 144, *145*

wanderer butterfly 178
wasps 179–183, *182*, *183*, 187–189, *187*, *188*, *189*
water beetles 112
water boatmen 96, *96*
water measurers 94, *95*
water scorpions *96*, 97
water striders 94, *94*
web spinners 77, *77*
weevils 125, *126*
whirligig beetles 112, *113*
whites (butterflies) 175
wings 14, *15*
Wingia aurata 159, 160
woodlice *10*, 11
wood moths 160, *161*

Xanthopimpla 184, *184*
Xenopsylla cheopis 133
Xixuthrus microcerus *124*, 125
Xylocopa aruana 196, 197

Zoraida 90, *90*
Zygaenidae 162, *163*
Zygoptera *47*, 48